T0295243

PRECISION FARMING AND PROTECTED CULTIVATION: CONCEPTS AND APPLICATIONS

The book consists of 32 chapters featuring the concepts and applications of precision farming and protected cultivation broadly covered with theoretical and practical approach. Chapter 1 to 8 are exclusively designed to provide detailed information on concept, need, objectives, benefits, components, applications and limitations of precision farming; laser leveler and its working mechanism, components and functioning; mechanized sowing and types of mechanical seeders and their use; approaches for mapping of soils and plant attributes; site-specific weed and nutrient management; precision management of insect-pests and diseases; yield mapping in horticultural crops.

An attempt has been made to cover the concept and application of protected cultivation in chapters from 9 to 30 characteristically highlighting the concept of greenhouse technology, its principles as well as historical and technological developments, classification of greenhouse structures and specifications for traditional/ low cost greenhouses, agrivoltaic system, its concept and features, response of plant species under greenhouse conditions, criteria for the selection of crops and varieties for protected cultivation, basic considerations for site selection, orientation and designing of greenhouse structures, climate control mechanisms for cooling and heating in greenhouses, components, accessories and BIS codes for protected cultivation, types of Irrigation system for greenhouse production system, growing media for greenhouse cultivation, soil pasteurization namely solarization, steam sterilization, chemical sterilization and augmentation with biological agents, checking the suitability of soil and water for greenhouse crops, plug tray nursery raising, basics of fertigation in greenhouse production system, packages of practice for greenhouse cucumber, bell pepper, tomato and melons, potential of pruning as unconventional alternative for mass multiplication of greenhouse cucumber and tomato, types of soil-less cultures, GAP for protected cultivation and economic analysis of protected cultivation.

Dr. Sanjeev Kumar is Assistant Professor (Vegetable Science) at ASPEE College of Horticulture & Forestry, Navsari Agricultural University, Navsari (Gujarat), India and has professional experience of 16 years.

Dr. Sanmukh N. Saravaiya, is serving as I/c Professor & Head, Department of Vegetable Science at ASPEE College of Horticulture and Forestry of Navsari Agricultural University, Navsari, Gujarat, India has a brilliant career throughout and has professional experience of more than 28 years.

Dr. Avnish Kumar Pandey is presently working as Assistant Professor (Horticulture: Fruit Science) at ASPEE College of Horticulture and Forestry, Navsari Agricultural University, Navsari, Gujarat. He has completed his B.Sc. (Hons.) Agriculture from CSAUA&T, Kanpur, (U.P.), M.Sc. Fruit Science (Hort.) from Dr. YSPUH&F, Solan, (H.P.) and received ICAR-JRF award and further completed Ph.D. in Fruit Science (Hort.) from NDUA&T, Ayodhya (U.P.).

PRECISION FARMING AND PROTECTED CULTIVATION: CONCEPTS AND APPLICATIONS

Sanjeev Kumar, S.N. Saravaiya and A.K. Pandey
Navsari Agricultural University
Navsari-396 450 (Gujarat), INDIA

NARENDRA PUBLISHING HOUSE
DELHI (INDIA)

First published 2021
by CRC Press
2 Park Square, Milton Park, Abingdon, Oxon, OX14 4RN

and by CRC Press
6000 Broken Sound Parkway NW, Suite 300, Boca Raton, FL 33487-2742

CRC Press is an imprint of Informa UK Limited

British Library Cataloguing-in-Publication Data
A catalogue record for this book is available from the British Library

Library of Congress Cataloging-in-Publication Data
A catalog record has been requested

ISBN: 978-1-032-05276-2 (hbk)
ISBN: 978-1-003-19684-6 (ebk)

Contents

NAVSARI AGRICULTURAL UNIVERSITY
NAVSARI 396 450 (GUJARAT) INDIA

Tel. (O): 02637-283869
Fax: 02637-282554
Email: vc@nau.in

Dr. S. R. CHAUDHARY
VICE CHANCELLOR

FOREWORD

The agriculture sector alone sustains the livelihood of around 64% of the population and contributes nearly 26% to the gross domestic product of India. There has been substantial increase in food grain production from about 51 million tonne in 1950 to about 291.95 million tonne in 2019-20. The worldwide production of horticulture has doubled over the past quarter century and the value of global trade in horticulture crops now exceeds that of cereals. Horticulture production surpassed the food production in India for the first time during 2013-14. India, though one of the biggest producers of agricultural products, has very low farm productivity. Average productivity of Indian farms is 33% that of the best farms world over. This productivity needs to be increased so that farmers can get more remuneration from the same piece of land with less labour. With increasing population, urbanization and contagious depletion of natural resources, there has to be a paradigm shift in perception from crop production to productivity and profitability. In the current scenario, the major challenges arising are shrinking land and depleting water as well as other related resources in agriculture.

There is a dire need for promoting farmer friendly location specific production and management strategies in a concerted manner to achieve vertical growth in horticulture duly ensuring quality of produce and better remuneration per unit of area with judicious use of natural resources. In this endeavour, precision farming aims to have efficient utilization of resources per unit of time and area for achieving targeted production of horticultural produce. Precision farming is a key component of the third wave of modern agricultural revolutions. One of the important applications of precision farming is protected cultivation, which has shown constant growth in India with the concerted efforts put forth by scientists and policy makers. Protected cultivation possesses numerous prospects to manage climatic diversity of varying levels of temperature, humidity, rainfall etc. effectively as compared to open cultivation and simultaneously fulfils the requirements of growers in terms of manifold increase in yield as well as quality of greenhouse crops.

I compliment Dr. Sanjeev Kumar, Dr. S.N. Saravaiya and Dr. A.K. Pandey for their prodigious endeavour in bringing out this publication "Precision Farming and protected cultivation: Concepts and Applications". The book emphasizes pedagogic information on concepts and applications of precision farming and protected cultivation and authors have tried to justify the contents of new course "Precision Farming and Protected Cultivation" recommended by 5th Deans' Committee with the help of good photographs and illustrations.

I am sure that this publication will serve as resource material not only to the undergraduate horticulture students but have significant use for PG students, researchers, academicians, stakeholders. I extend my appreciation and good wishes to the authors for their feat in writing this book.

Navsari, India

(S.R. Chaudhary)

Preface

The present day cultivation practices in various crops are largely targeted at using water resources injudiciously. A significant change in climate on a global scale will impact agriculture and consequently affect the world's food supply. Climate change per se is not necessarily harmful; the problems arise from extreme events that are difficult to predict. More erratic rainfall patterns and unpredictable high temperature spells will consequently reduce crop productivity.

Precision farming proposes to prescribe tailor made management practices, as it recommends application of right amount of inputs at right time and location to achieve higher productivity and sustainable development, optimize production efficiency and product quality increase input use efficiency, manage pests efficiently, conserve natural resources with minimum environmental impact and risk. Protected cultivation- an important precision farming application, have shown steady growth and rapid expansion in India and worldwide. The phase-wise implementation of protected cultivation in India through financial assistance and mass awareness created about the technology has given a stimulus for better acceptance and adoption of this technology.

It highlights the concepts and applications of precision farming and protected cultivation comprehensively covered with theoretical and practical approach. The publication presents detailed information on concept, components, applications of precision farming, working mechanism of laser leveler, mechanized sowing, mapping of soil, plant attributes including yield, precision management of inputs and pests.

We are highly indebted to the authors of different earlier books, research articles, web sites, links whose vision and technical support in respect of the subject have helped us in shaping up this material in a book form with our experience and research on some of important aspects.

We are also thankful to Navsari Agricultural University for invariable motivation in the preparation of this text. We proclaim our ecstatic gratitude to Dr. S.R. Chaudhary, Director of Research & Dean PG Studies, NAU and Dr. P.K. Shrivastava, Principal & Dean, ACHF, NAU, Navsari for being kind and considerate, whose constant encouragement and constructive criticism have motivated us to bring this manuscript to the present form. Finally, we owe the obvious support and sacrifice of our families. It would be unrealistic to suppose that the book is free from errors and the suggestions will be highly appreciated for making improvement in this book.

Authors

1

PRECISION FARMING: CONCEPT, NEED, OBJECTIVES AND BENEFITS

Background Information

- Agriculture with its allied activities is the largest sector in India revealing 70% dependency of its rural households for their livelihood and has played important roles for food and nutritional security, poverty alleviation, employment generation and sustainable development.
- Indian is the largest producer of cotton, pulses and milk, second largest producer in food grains, fruits and vegetables, and third in egg production.

  Agriculture sector alone contributes about 14.5% towards Gross Domestic Product (GDP) of India.
- Food grain production has experienced considerably a significant jump from about 51 million tonne during 1950 to 291.95 million tonne in 2019-20. The production of horticulture products has doubled over the past quarter century and the value of global trade in horticulture crops now exceeds that of cereals. For the first time, horticulture production surpassed food production in India during 2013-14 and continues to excel currently.
- India is characteristically a country of small agricultural farms, where approximately 80% of total land holdings in the country are less than 2 ha with 30% irrigated land only.
- India produces a wide variety of agricultural products because of its varied agro-climatic regions but low farm productivity is major concern, which is around 33% of the best agricultural farms world over. Indian farmers can get more remuneration from the same piece of land with fewer inputs, once the productivity gets increased.
- Asia and the Pacific region accounts for more than 70% of the global agricultural population but has only 30% of the world's arable land. An increase in yield of these regions over the years has largely been due to excessive and

indiscriminate use of inputs like irrigation, seeds, fertilizers, pesticides, agrochemicals *etc.* and at considerable expense of its natural resources.

- Soil erosion in this region due to water and wind exceeds the natural soil formation by 30-40 folds. The problem of water quality deterioration is also very serious. Pollution of drinking water in tobacco and rice ecosystems of Malaysia, and of groundwater near vegetable fields in Japan are just two examples. However, the increasing rate of population demands much higher pace in the increase of food production on one side, and needs utmost attention in maintaining harmony with the environment-the core concept of sustainable agriculture.
- Most of the pesticidal recommendations are also region specific, which are on a few random observations of insect-pests and diseases density/severity. Because of substandard quality of agrochemicals often due to adulteration and increased pest resistance, indiscriminate application is now common especially in high valued crops.
- Comprehensive and reliable information on land use, soils (extent of waste lands and degraded lands), agricultural crops, water resources (both surface and underground), natural hazards/ calamities like drought and flood and agro meteorology is essential. Season-wise information on crops, their acreage, vigour and production enables the country to adopt suitable measures to meet shortages if any and implement proper support and procurement policies.
- With increasing population, urbanization and over exploitation of natural resources, there has to be a paradigm shift in farmers' perception from production towards productivity and profitability. At present, agriculture is facing major challenges due to shrinking land holding, depleting water and other related resources. There is an urgent need for adopting farmer friendly location-specific production and management strategies in a concerted manner to achieve vertical growth in horticulture production with ensured quality of produce and judicious use of natural resources for better return per unit area. In this context, precision farming has potential of utilizing resources per unit of time and area efficiently for achieving targeted production in horticultural crops.
- Precision agriculture is proposed as a key component of the third wave of modern agricultural revolutions. The first agricultural revolution came with the advent of increased mechanization during the period from 1900 to 1930s and each farmer produced enough food to feed about 26 people during this time. The 1960s prompted the "Green Revolution" with new methods of genetic modification, which led to each farmer feeding about 155 people. It is expected that the global population will reach about 9.6 billion by 2050 and

food production must effectively double from current levels in order to feed every mouth. With new technological advances in agriculture through precision farming, each farmer will be able to feed 265 people from the same area.

Concept of Precision Farming

Precision Agriculture (PA) also known as precision farming, prescription farming, variable rate technology (VRT), site-specific farming (SSF), site-specific management (SSM), site-specific crop management (SSCM), is considered as the vibrant agricultural system of the 21st century, as it symbolizes a better balance between reliance on traditional knowledge as well as information and management intensive technologies.

Professor Pierre C. Robert, who is considered as the father of precision farming defined precision farming as precision agriculture which is not just the injection of new technologies but it is rather an information revolution, made possible by new technologies that result in a higher level, a more precise farm management system. **It basically means adding the right amount of input at the right time and the right location within a field. It's all About managing specific sites following the fundamentals of 3 R.**

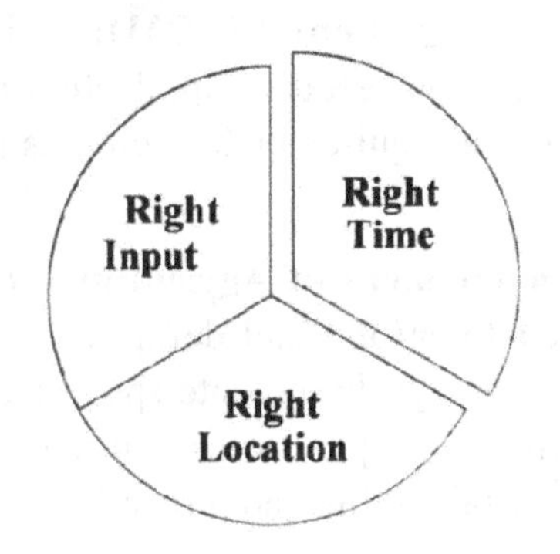

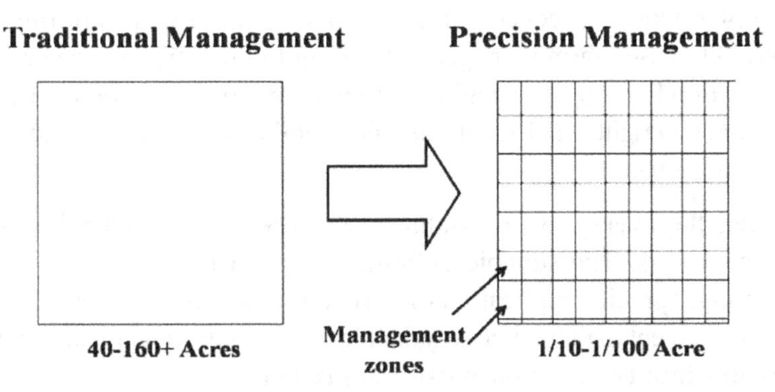

Precision Farming and its approach

Precision agriculture (PA) ("precision farming", "site-specific farming", "farming by the foot", "spatially variable crop production", "grid farming"), is an innovative conception of agricultural production based on information technologies in crop production.

Precision Agriculture is the application of technologies and principles to manage spatial (space-related/geographic) and temporal (time related) variability associated with all aspects of agricultural production for improving production with environmental safety.

Precision Agriculture can also be defined as a comprehensive system designed to optimize agricultural production through the application of crop information, advanced technologies and management practices.

Definitions of Precision Agriculture Given by US House of Representatives, 1997

Precision Agriculture (PA) is integrated information and production based farming system that is designed to increase long term, site specific and whole farm production efficiency, productivity and profitability, while minimizing unintended impacts on wildlife and the environment.

Site-specific crop management (SSCM): It is a form of precision agriculture whereby decisions on resource application and agronomic practices are improved to fit well to the requirements of crop as per soil heterogeneity of the field.

The United States Department of Agriculture (USDA) refers this kind of agriculture "as needed farming" and define it as "A management system that is information and technology based, site specific and uses one or more of different sources of data like soil types, crops, nutrients, pests, moisture or yield for optimum profitability, sustainability and protection of the environment".

In simpler words, precision farming can be defined as information and technology (IT) based farm management system to identify, analyze and manage variability within fields by doing all the practices of crop production in right place, at right time and in right way for optimum profitability, sustainability and protection of the land resources.

In India, the average land holdings are very small, even with large and progressive farmers. The suitable definition of precision farming in context to Indian farming, "precise application of agricultural inputs based on soil, weather and crop requirement to maximize sustainable productivity, quality and profitability *i.e.* **minimum input-maximum output approach**.

Rationale of Precision Farming

The "Green revolution" (1960s) has made our country self-sufficient in food production attributable to High Yielding Varieties (HYVs), fertilization, irrigation, pesticides and increase in cropping intensity (CI) and lastly mechanization of agriculture. However, Green Revolution Technologies (GRTs) has also led to some challenges, which needs to be addressed for increasing productivity of crops and simultaneously safeguarding ecological interest.

1. **Fatigue of Green Revolution:** No doubt, India has experienced spectacular growth in agriculture, however the productivity of many major crops are still below the expected levels. India has not achieved even the lowest level of potential productivity of Indian high yielding varieties, while the world's highest productive country has crop yield levels significantly higher than the upper potential of Indian HYVs.
2. **Natural Resource Degradation:** The green revolution has also been associated with some negative factors leading to ecological imbalance. Indian environment statistics shows that about 182 million ha of the country's total geographical area of 328.7 million ha is affected by land degradation, of which large proportion of 141.33 million ha is due to water erosion, 11.50 million ha due to wind erosion, and 12.63 and 13.24 million ha are due to water logging and chemical deterioration (salinity and loss of nutrients), respectively. On the other hand, India shares 17% of world's population, 1% of gross world product, 4% of world carbon emission, 3.6% of CO_2 emission intensity and 2% of world forest area. Hence, there is a dire need to convert this green revolution into an evergreen revolution with farming systems approach that can help to produce more from the available land, water and labour resources without ecological and social imbalance.

Therefore, it is important to learn the critical differences between traditional and precision farming in order to get basic idea on realistic figures leading to better understanding and adoption of new applications of precision farming.

Traditional farming VS.	Precision farming
Arable site is considered as a homogenous for applying treatments/inputs at field level.	Arable site is regarded as different from one point to another and considered as heterogeneous at field level.
Nutrient management is based on representative samples taken from a field.	Nutrient management is based on point like samples (site-specific) taken at specific sites with the use of GPS.
Plant protection applications for diseases and insect-pests are generally made based on average survey done for plant damage.	Plant protection applications are made based on GPS and point like plant survey (specific survey).

[Table Contd.

Contd. Table]

Traditional farming	VS.	Precision farming
Selection and planting of crops and variety (ies) is limited to overall arable site.		Crops and variety (ies) are selected as per the suitability to a specific site/location.
Farm operations are adhered to same machinery in the arable site.		Farm operations by machinery are adjusted specifically to the arable site.
Plant stocks are minimally organized into homogeneous blocks at arable sites, hence scope for better management is less.		Unified plant stocks are organized into homogeneous blocks at arable sites for better management.
Few data are available influencing decision making at arable site.		A lot of data are available to make conclusive decisions for better growth and development of a crop.

The critical differences between traditional and precision faming clearly define the prospects of precision farming for sustainable development of agriculture/ horticulture in India.

Objectives of Precision Farming

Higher profitability and sustainability: Maximum profit can be achieved in each zone or site of a field by optimizing precise application of inputs like variety, seed, fertilizer, herbicide, pesticide *etc.* as per crop demand, which can be determined by weather, soil characteristics (nutrient availability, texture and drainage *etc.*) and historic crop performance.

Optimizing production efficiency: Identification of variability in yield potential may offer possibilities to optimize production quantity at each site or within each zone using differential approaches under given set of field conditions.

Increasing efficiency of inputs: Efficient use of inputs like fertilizers, seeds *etc.* according to the yield potential of soil at a given location.

Effective and efficient pest management: Minimizing inputs cost for crop production is one of the important objectives of precision farming, which can result in getting higher return and better environmental services. Site-specific variable rate application recommends the application of chemicals *i.e.* herbicides, pesticides at the area of problem with targeted approach in comparison to conventional farming methods.

Optimizing product quality: Product quality can be optimized by using sensors for the detection of quality attributes of a crop, which helps in taking decision on input application as per the target.

Conservation of soil, water and energy: A comprehensive approach under precision farming begins from crop planning, assessing field variability, which includes those tillage practices which disturb the soil to its minimum level. In addition to this, water is efficiently applied by using techniques like drip irrigation, sprinkler irrigation *etc.* with the principle of more crop per drop. In all such precision applications, very less energy is used thus leading to conservation of energy also.

Protection of surface and ground water: Safeguarding the environment by way of efficient use of inputs like fertilizers, chemicals etc. which prevents their leaching through ground water or as runoff.

Minimizing environmental impact: Better management decisions under precision farming are made to modify inputs for meeting out the production needs, which ensures no or negligible loss of any applied input to the environment.

Minimizing risk: Risk management is a common strategy being adopted today by most of the farmers, which can be anticipated as income and environmental safety. In a production system, farmers often practice risk management by misjudging on the side of extra inputs, while the unit cost of a particular input is adjudged 'low'. Thus, a farmer may put an extra spray on, add extra fertilizer, buy more machinery or hire farm extra labour to ensure that the product is produced/ harvested/sold well in time thereby ensuring economic return.

Prospectives of Precision Farming

The effectiveness of precision farming is highly dependent on two factors: 1) Existence and assessment of variability within the fields/farms 2) Ability of a producer/farmer to identify and put into use the best management practices for each field's sub-area based on variability assessment. If the spatial data provided by precision farming is properly used, the following potential benefits can be realized:

Increased profits through increased efficiency: A producer can use agricultural inputs more efficiently as per the variability present for those components in specific sites of a field by applying more or less of them depending upon the specific requirements.

Reduced agronomic inputs: Producers/farmers can reduce their overall use of agricultural inputs like fertilizers, chemicals etc. by making adjustment to a field's fertility level based on assessment, thus the amount of inputs applied to a specific site can specifically be applied.

Better record keeping: A large amounts of data on spatial records of inputs and outputs for agricultural fields generated by precision farming can help to develop more accurate management plans and strategies.

Improved production decisions: Precision farming can effectively be used to make decisions about land use. Profit maps of a farm showing the spatial distribution of a field's profitability can help a producer to make decisions about adoption of cropping systems at the best.

On-farm research: The ability to quantify the spatial performance of crops allows conducting more comparison trials within fields of a location. For instance, it is relatively simple to compare the performance of different varieties on different soil types, when practicing precision farming.

Reduced environmental impact: Production inputs are applied and used more efficiently under precision farming, so very low or minimum fraction of such inputs will leave the field through surface water and ground water. Reduced environmental impact will certainly have impact on the acceptance of farm produce in near future.

Property advantages: Many landlords are giving preference to farmers who can create yield maps and other files of spatial data for the fields under production practice. This spatial history may also increase the value of crop land.

More area farmed: The records generated under precision farming can allow managing more crop land effectively than the effectiveness observed in the past.

2

PRECISION FARMING: COMPONENTS, APPLICATIONS AND LIMITATIONS

Precision Farming is a holistic farm management approach which allows farmers to adjust inputs use and cultivation practices including variety selection, seed, sowing/planting, fertilizers, pesticides and irrigation water, tillage, harvesting harmonizing to varying soil texture, fertility and other field characteristics. Precision farming differs from conventional farming which is mostly based on uniform applications and treatments all across a field, while PF involves mapping and analysis of field variability which links spatial relationships to management tasks, thus permitting farmers to view and manage their farms, crops and practices with a new perspective. Thus, PF has ability to provide a network of information for production and management decisions at field level.

Elements/Components of Precision Farming

A. Information or data base

Soil: Texture, Structure, Physical condition, Moisture, Nutrients *etc*.

Crop: Plant Population, Plant Tissue Nutrient Analysis Status, Crop Stress, Weed patches (weed type and intensity), Insect or fungal infestation (species and intensity), Crop yield, Harvest Swath Width *etc*.

Climate: Temperature, humidity, rainfall, solar radiation, wind velocity *etc*.

Field variability (spatially or temporally), soil related properties, crop characteristics, weed and insect-pests population, and harvest data are important data bases for realizing the potential of precision farming.

B. Technology

Precision farming deals primarily with understanding the variability of field and its management over time and space. Precision farming includes a variety of tools of hardware, software and equipments like Global Positioning System (GPS), Differential Global Positioning System (DGPS), Geographic information systems (GIS), Remote sensing, Sensors, Variable Rate Applicator, yield mapping technology/yield monitors, Computer system *etc.* for variability assessment and its management in the field.

1. **Global Positioning System (GPS):** GPS makes available continuous position (location) information in real time, while in motion. Positioning system works with the help of different constellations of satellites and developments of such different positioning systems are the main technological interventions bringing precision farming concept into reality. GPS receivers can directly be carried to the field or mounted on implements that allow users to return to specific locations to sample or treat such areas and helps to allow mapping of soil and crop measurements at any time. GPS provides accurate positioning system for field implementation of variable rate technology. It helps to manage application of inputs by equipments and identify the precise location of farm equipment within specific positions of the field for precision application of fertilizers and pesticides to the soil-plant characteristics. GPS receiver with electronic yield monitors is generally used to collect yield data across the land in precise way. Global positioning systems (GPS) are widely used for mapping yields (GPS + combine yield monitor), variable rate planting (GPS + variable rate planting system), variable rate fertilizer application (GPS + variable rate controller), field mapping for records and insurance purposes (GPS + mapping software) and parallel swathing (GPS + navigation tool).

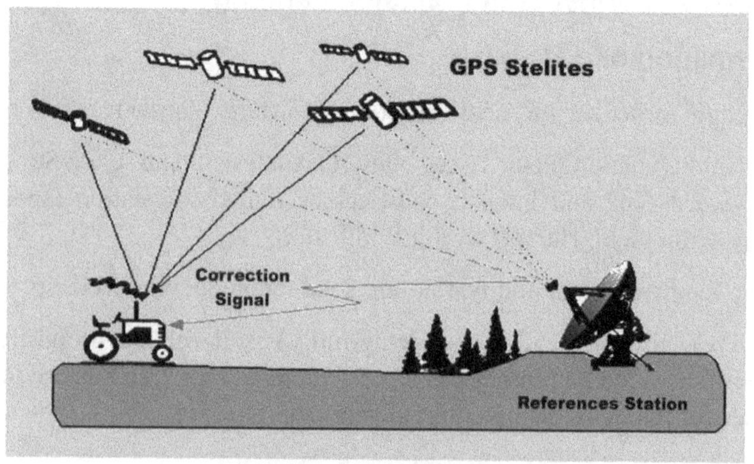

Global Positioning System

2. **Differential Global Positioning System (DGPS):** Differential Global Positioning System is a technique to improve GPS accuracy that uses pseudo range errors measured at a known location to improve the measurements made by other GPS receivers within the same geographic area. DGPS makes use of a series of satellites that identify the location of farm equipment within inches of an actual site in the field. The knowledge of a precise location helps to compare different locations of soil samples and the laboratory results in a soil map, thus enable to prescribe fertilizers and pesticides to fit well to soil properties like clay and organic matter content and soil conditions (relief and drainage). Various tillage adjustments can be made with the help of information generated though this system. The system also allows monitoring and recording yield data as one goes across the field. The systematic working of DGPS is shown in picture below.

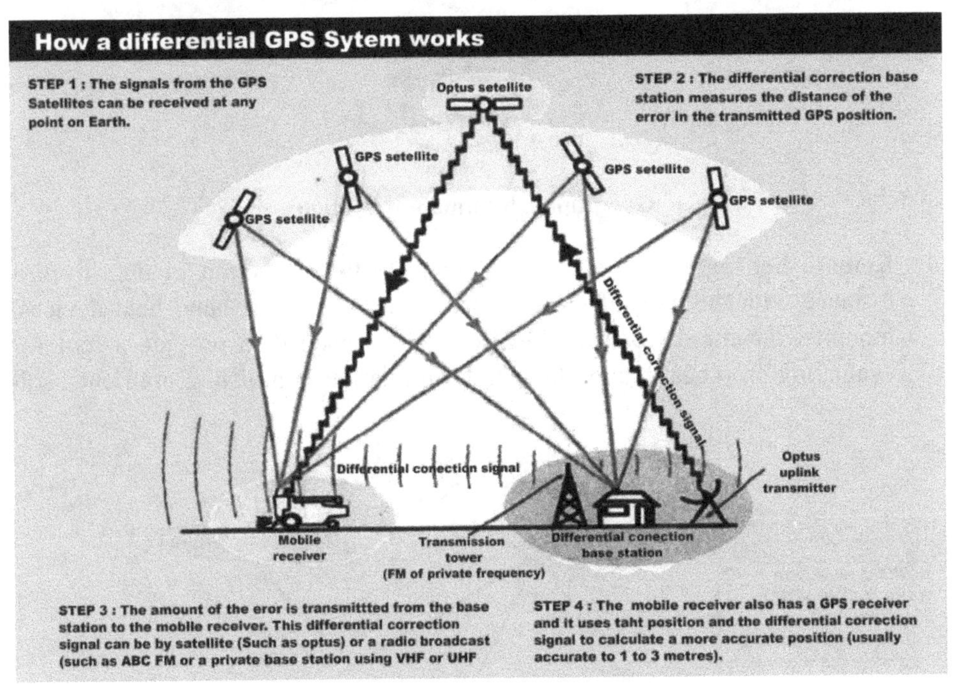

Differential Global Positioning System

3. **Geographic Information System (GIS):** Geographic information systems (GIS) are computer hardware and software with featured attributes and location data to produce maps. GIS includes organized collection of computer hardware, software, geographic data and personal designed to capture, store, update, manipulate, analyze and display all forms of geographically referenced information efficiently. Agricultural GIS stores a series of information such as yield, soil survey maps, remotely sensed data, crop scouting reports and soil

nutrient levels. GIS contains base maps like topography, soil type, nutrient level, soil moisture, pH, fertility, weed and pest intensity and can also integrate all types of information and interface with other decision support tools for application of recommended rates of nutrients or pesticides.

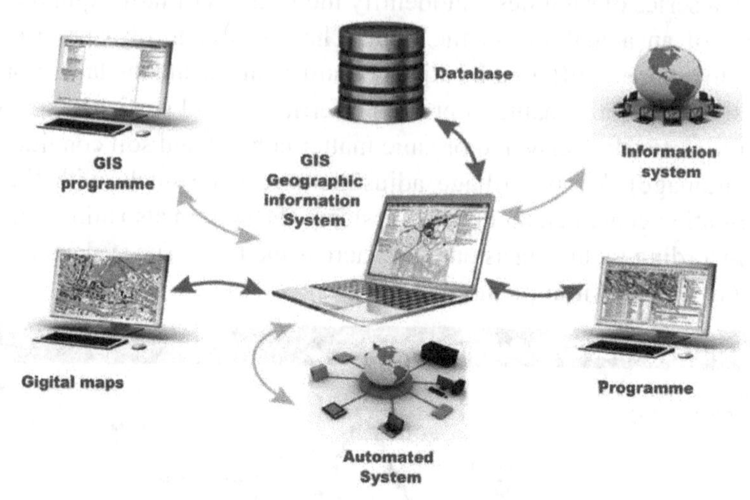

Geographic Information System

4. **Remote Sensing:** Remote sensing refers to the collection of data from a distance with the help of sensors, which can simply be hand held devices, mounted on aircraft or satellite. Remotely sensed data provide a tool for evaluating crop health and plant stress in relation to moisture, nutrients, soil

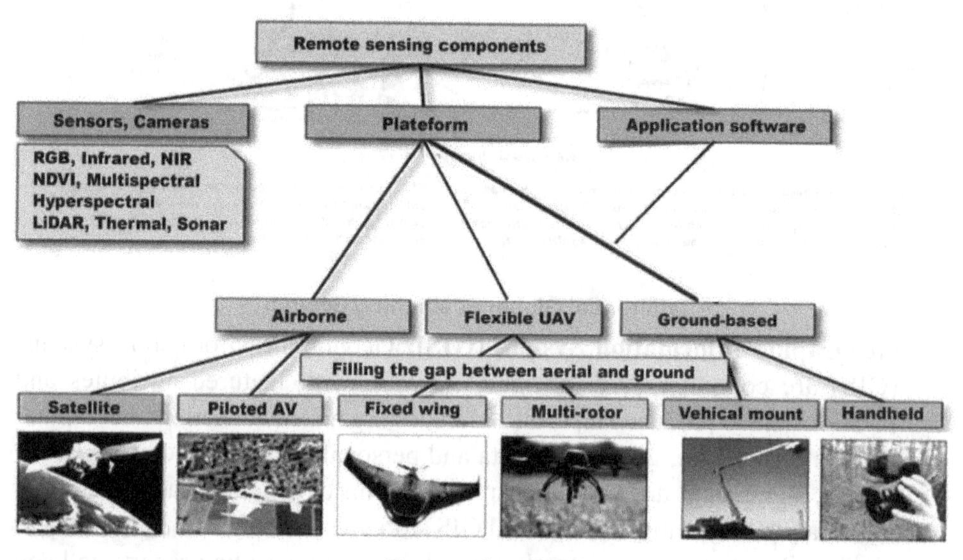

Remote Sensing

compactness, plant insect-pests and diseases, and other plant health concerns oftenly and easily detected in overhead images. Remote sensing can reveal and analyze seasonal variability affecting crop yield and can be helpful to make timely management decisions to improve profitability for the current crop. Remotely sensed data obtained either by aircraft or satellite containing electromagnetic emittance and reflectance data of a crop can provide information useful for soil condition, plant growth, weed infestation *etc.* This type of information is cost effective and can be very useful for site-specific crop management programmes.

Developments on remote sensing by India: Crop production forecasts using satellite remote sensing data was conceptualized by Indian Space Research Organization (ISRO) in early eighties. IRS-1A, IRS-IIB, IRS-IIIC, IRS-1D are the Indigenous satellites used in remote sensing for agricultural purposes. IRS-P6/RESOURCESAT-1 is the tenth satellite of ISRO in IRS series, intended to not only continue the remote sensing data services provided by IRS-1C and IRS-1D, both of which have far outlived their designed mission lives, but also to vastly enhance the data quality. RESOURCESAT-1 is the most advanced Remote Sensing Satellite built by ISRO as of 2003.

This led to the success of CAPE (Crop Acreage and Production Estimation) project that was done with active participation of Ministry of Agriculture and Farmers' Welfare (MoA&FW) towards forecasting of crop production in selected regions. In order to enhance the scope of this project, the FASAL (Forecasting Agricultural Output using Space, Agro-meteorology and Land based Observations) programme was launched by developing methodology for multiple in-season forecasts of crops at national scale. A centre named Mahalanobis National Crop Forecast Centre (MNCFC) was established by MoA&FW in April 2012 at New Delhi, which operationally uses space-based observations at national level for pre-harvest multiple crop production forecasts of nine field crops namely wheat, rice, jute, mustard, cotton, sugarcane, *rabi* & *kharif* rice and rabi sorghum. Remote Sensing based acreage and yield forecasts based on weather parameters or spectral indices are used to provide production forecasts. The center is also actively involved in national level assessment of horticultural crops and their coverage across the agro-climatic regions of the country. RISAT-2BR1 is a synthetic-Aperture Radar (SAR) imaging satellite for reconnaissance built by ISRO. It is part of India's Radar Imaging Satellite (RISAT) series of SAR imaging spacecrafts and fourth satellite in the series and meant to provide services in the field of Agriculture, Forestry and Disaster Management.

In addition, the center also uses the space based techniques in making multiple assessments of drought conditions in the country, which in turn helps the Government in making decisions on relief measures for the affected areas in the country. A well-established Agriculture Drought assessment mechanism is used operationally by the MNCFC.

Considering the importance of horticulture towards food and nutritional security including its export potential and economic benefits, a programme called "Coordinated programme on Horticulture Assessment and Management using geoiNformatics (CHAMAN)" was launched in September 2014 by MoA&FW. The components of CHAMAN are a) Crop Inventory : 7 major horticultural crops in selected districts of major states (185 districts in 12 states), b) Development and Management Planning: Post-Harvest Infrastructure, Aqua-horticulture, Orchard rejuvenation, Crop Intensification, GIS Database creation, site suitability assessment and c) Research & Development: Crop identification, yield modelling and disease assessment, precision farming, new techniques and algorithms.

The hallmark of Indian space programme is the application-oriented focus and the benefits that have accrued to the country through these programmes. The societal services offered by Earth Observation, SATCOM and the recent NavIC constellation of satellites in various areas of national development. During past many years, Indian Remote Sensing Satellite constellation has taken giant strides in ensuring many areas of application, operational. Some of the most prominent ones are **Agricultural Crops Inventory, Water Resources Information System**, Ground Water Prospects, Forest Working Plans, Biodiversity and Coral Mapping, Potential Fishing Zones, Ocean State Forecasts, Rural Development, Urban Development, Inventory & Monitoring of Glacial Lakes / Water Bodies, Location based Services using NavIC constellation, Disaster Management Support Programme (Cyclone and Floods Mapping & Monitoring, Landslide Mapping & Monitoring, Agricultural Drought, Forest Fire, Earthquakes, Extreme Weather Monitoring and experimental Forecasts *etc.*).

Web Geoportals and mobile technologies (Bhuvan Geoportal) are the other popular platforms, being used by Governments, to provide information services and solutions at all levels, which are proving to be effective. The Government system has successfully adopted to use such technologies for the benefit of people at large.

5. **Sensors:** Remote sensors are generally categorized as aerial or satellite sensors that can provide instant maps of field characteristics. An aerial photograph is an example of optical sensor that can show variations in field colour corresponding to the changes in soil type, crop development, field boundaries, roads, water *etc.* Both aerial and satellite imagery can also be processed to provide vegetative indices reflecting plant health. Sensors are used to determine

crop stress, soil properties, pest incidence *etc*. These can also be used to measure soil and crop properties as the tractor passes over the field, as a scout goes over the field on foot or as an airplane or satellite photographs the field from sky. Yield monitors are the primary sensing system that makes measurements. Commercial sensing systems are available that claim to measure soil properties and make changes in application rate.

Sensors can also be carried by a scout to the field and used to sense the health of plants and soil properties. These use light reflectance on leaves to determine chlorophyll levels, which directly correlate nitrogen levels in the plant proportionally to the chlorophyll production. Several soil chemistry mini kits sensors are also available to measure soil pH, EC, nitrogen, potassium, phosphorus *etc*. directly in the field.

Application of Sensors in Agriculture

6. **Variable Rate Applicator:** The variable rate applicator has three components namely computer controller, locator and actuator. A computer system mounted on a variable rate applicator uses the application map and a GPS receiver to direct a product delivery system that changes the amount and/or kind of product according to the application map *e.g.* Combine harvesters with yield monitors. Two methods are generally used for variable rate applications *viz*., first one is map based on historical data (previous or present year) that help the process control technologies to draw information from the GIS (prescription maps) and adjust fertilizer application, seeding and pesticide selection and application rate, thus providing proper management of the inputs. The second method

uses sensors that can adjust the applications rates of equipments by detecting some characteristics of plants or soil.

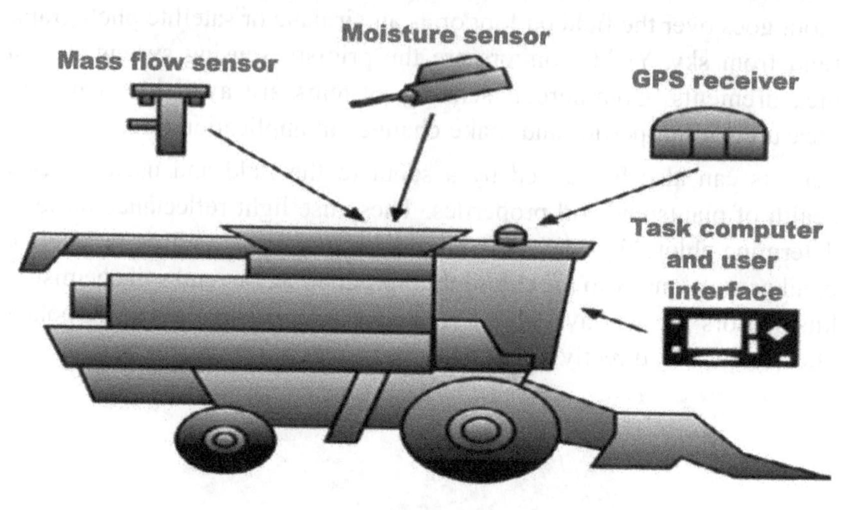

Yield mapping technology

7. **Yield Mapping Technology:** Yield is an ultimate indicator of variations of different agronomic/horticultural parameters in different parts within a field. So, mapping of yield and its interpretation, and correlation of this map with the spatial and temporal variability of different crop parameters help in developing management strategy for successive season of a crop. Present yield monitors measure the volume or mass flow rate to generate periodical record of quantity of crop harvested during that time period and then, periodic yield data is synchronized with location obtained from onboard GPS system to create most common colour coded thematic map. Yield mapping can be carried out easily in mechanized crops, for example, loading cells weigh the crop passing on a conveying belt or an array of sonic beam mounted over the grape discharge chute to estimate the volume and the tonnage of fruit harvested. Radio-frequency identification (RFID) or bar code tags on the bins have also been used for yield mapping in which a weighing machine is combined with a tag reader and a GPS to record the weight and the place of each bin. The data collected are used to produce yield maps of the orchard.
8. **Computer System:** Precision farming requires the acquisition, management, analysis and output of large amount of spatial and temporal data generated for a specific field. Computer software has now become better with time and the knowledge needed for managing variability and decision making on the field can be generated via such softwares to execute management strategies.

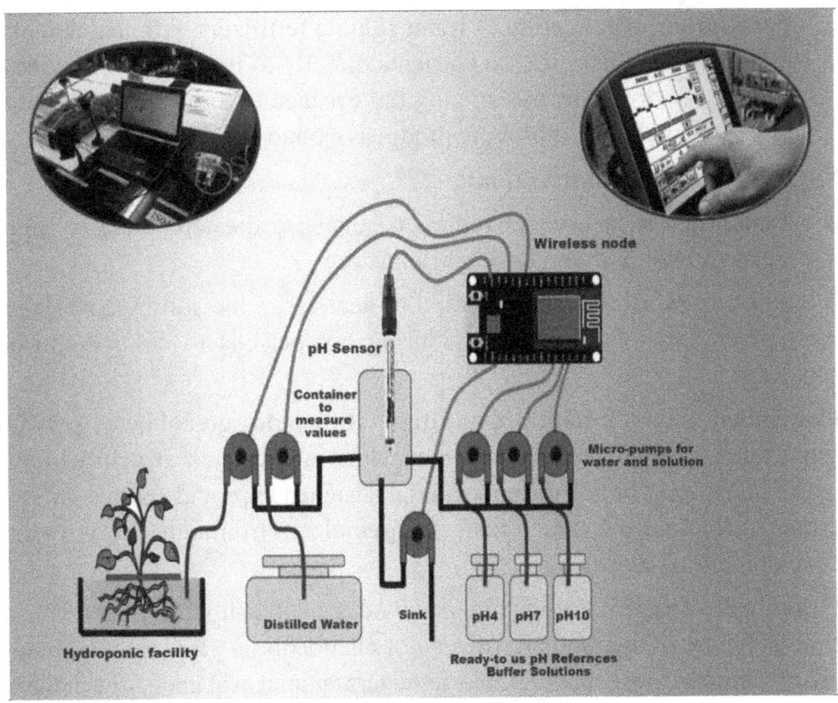

Computer System

Applications of Precision Farming

1. **Water Management:** Drip irrigation, sprinkler irrigation and fertigation are the important applications in precision farming for water management.

Drip Irrigation   Fertigation   Sprinkler Irrigation

   a. **Drip Irrigation:** It is a regulated or slow application of irrigation water through emitters at frequent interval around crop rhizosphere over a period of time. It saves about 50-70% water, reduces labour and energy cost and has almost negligible weeds problem.

   b. **Sprinkler Irrigation**: Water is sprayed into the air and allowed to fall on the ground surface somewhat resembling to rainfall. The spray is regulated by the flow of water under pressure through small nozzles.

c. **Fertigation:** Application of water soluble fertilizers with irrigation water to the crop, which provide nutrients directly to the active root zone, less likely to be lost thus minimizing the wastage of expensive nutrients and ultimately helps in improving productivity and quality of produce.

2. **Surface Covered Cultivation:**

a. **Mulching:** It is used to reduce or increase the temperature, suppress weed growth and conserve soil moisture.

b. **Soil Solarization:** It is a method of heating up the soil by covering with transparent polythene sheets during hot periods to control soil-borne pathogens.

3. **Controlled Environmental Condition**: It includes greenhouse, polyhouse, grow tunnels, shade net houses *etc.* which are framed structures covered with a transparent or translucent material designed to provide protection against climatic fluctuations and create congenial environment for growth and development of crops.

4. **Organic Farming:** It uses natural and biodegradable inputs which deliberately avoids the use of synthetic fertilizers/ or chemicals *etc.* It mainly includes the use of vermicompost, manures, bio-fertilizers, animal husbandry, green manures, biological management and crop rotation *etc.*

Mulching

Soil Solarization

Controlled Environmental

Organic Farming

5. **Precise Space Utilization:**
   a. **High Density Planting (HDP):** HDP is a system of planting more number of plants than the optimum or normal through manipulation of tree architecture. Different methods like genetically dwarf scion cultivars, dwarfing rootstock, training and pruning, use of growth retardants are generally adopted to manage the size of plants to fit well under HDP.
   b. **Meadow Orcharding:** It is also known as ultra high density planting system which accommodates 20000-100000 plants/ha which is literally called as grassland. In order to maintain tree form, severe top pruning is practiced similar to mowing of grassland.

High Density Planting (HDP)

Meadow Orcharding

6. **Micro Propagation**: It refers to the production of plant from very small plant parts, tissues or cells (explants) grown aseptically in a test tube or containers under controlled environment.
7. **Integrated Pest Management (IPM)**: A pest management strategy which involves different components such as use of resistant varieties, managing the natural predators, adopting cultural practices, physical, mechanical, biological, curative methods and judicious application of pesticides to minimize

pest population to a state below threshold level thereby avoiding economic injury to plants.

8. **Integrated Nutrient Management (INM):** It involves application of nutrients from organic as well as inorganic sources with the aim of improving physico-chemical and biological properties of the soil.

9. **Vegetable Grafting**: It is a technique of uniting two plants, one which is horticulturally superior and has susceptibility or less tolerance against biotic/ abiotic stresses, is used as scion and second with tolerance/resistance against biotic/ abiotic stresses is used rootstock. Though, improvement work in vegetables crops has produced so many horticulturally superior varieties in the country and worldwide, however production constricts like soil-borne pathogens are difficult to address through breeding approaches owing to the complexity of inheritance for such traits. Sometimes, crossing barriers between different sources of resistance or tolerance make it impossible to transfer desirable genes into a crop of interest. Under such situations, vegetable grafting proves to be a potential alternative technique to mitigate various biotic and abiotic problems like nematodes, root rots, salinity, flooding *etc*.

Limitations of Precision Farming under Indian Conditions

- Small sized farms/ small land holdings
- Presence of heterogeneous cropping systems
- Precision farming technology demands high initial capital investment
- High cost of obtaining and analyzing site-specific data
- Complexity of precision tools and techniques requiring new technical skills
- Precision farming as a new technology needs to be narrated and demonstrated to farmers with its on-farm and off-farm impacts on yields for better adoption
- Lack of local technical expertise
- Uncertainty on returns from investments on new equipments and information management system
- Culture and attitude including reluctance of farmers towards adoption of new technologies and lack of awareness about environmental problems
- Knowledge and technological gaps

3

LASER LEVELER: WORKING MECHANISM, ITS COMPONENTS AND FUNCTIONING

Land leveling is a precursor to good agronomic, soil and crop management practices. Resource conserving technologies perform better on well leveled and laid out fields. Effective land leveling is meant to optimize water use efficiency, improve crop establishment, reduce weed problem, improve uniformity of crop maturity, reduce the irrigation interval and effort required to manage crop with significant effect on yield. The benefits of effective leveling have been well recognized, so considerable attention and resources are involved in leveling of fields properly. There are two types of methods used for leveling the land; manual and mechanized.

Traditional Land leveling

Tractor drawn levelers

Laser leveler

Traditionally farmers level their fields using animal drawn or tractor-drawn levelers. These levelers are implements consisting of a blade acting as a small bucket for shifting the soil from higher to the low-lying positions. It is seen that even the best leveled fields using traditional land leveling practices are not precisely levelled and this leads to uneven distribution of irrigation water. The common practices of irrigation in intensively cultivated irrigated areas are flood basin and check basin irrigation systems. These practices on traditionally levelled or unlevelled lands lead to water logging conditions in low-lying areas and soil water deficit at higher spots. The advanced method to level or grade the field is to use laser-guided leveling equipment. Laser land leveling refers to leveling of field within certain degree of desired slope using a guided laser beam throughout the field.

Benefits of Precise Land Leveling

1. Saves irrigation water by more than 35%
2. Reduced weed problems in the field
3. Helps in increasing net cropped area by about 3.5%
4. Reduce farm operating time by 10%
5. Assist top soil management
6. Saves labour costs
7. Saves fuel/electricity used in irrigation
8. Increase productivity up to 50%
9. Increase in water application efficiency up to 50%
10. Reduces the amount of water required for land preparation (20-30%), hence saving in energy (diesel/electricity)
11. Better crop stand due to even application of fertilizers and other inputs *i.e.* improvement in crop yield by 10 to 15%
12. Totally automatic (less load on operator)
13. Effective land leveling reduces the work in crop establishment and crop management, and increases the yield and quality
14. Improves uniformity of crop maturity.
15. Decreases the time to complete tasks.
16. More uniform distribution of water in the field.
17. More uniform moisture environment for crops.
18. More uniform seed germination and fast growth of crops.
19. Reduction in seeds rates, fertilizer, chemicals and fuel used in cultural operations.

Laser Leveling is also Associated with Few Limitations

1. High cost of the equipments/laser instruments.
2. Requires skilled personals to set/adjust laser settings and operate the tractor.

Working Mechanism of Laser Leveler

The system includes a laser transmitting unit which emits an infrared beam of light that can travel up to 700 m in a perfectly straight line. The second part of the laser system is a receiver that senses the infrared beam of light and converts it to an electrical signal. The electrical signal is then directed by a control box to activate an electric hydraulic valve. Several times a second, this hydraulic valve raises and lowers the blade of a grader to keep it following the infrared beam. Laser leveling of a field is accomplished with a dual slope laser that automatically controls the blade of the land leveler to grade the surface precisely in order to eliminate all undulations tending to hold water. Laser transmitter creates a reference plane over the work area by rotating the laser beam 360°. The receiving system detects the beam and automatically guides the machine to maintain proper grade. The land can be leveled or sloped in two directions and this is accomplished automatically without the operator touching the hydraulic controls.

Major Components and Connectivity of a Laser Leveling System

Tractor: A 4 wheel tractor is required to drag the leveling bucket. The size of the tractor can vary from 30-500 hp depending on the time restraints and field sizes. In Asia, tractors ranging in size from 30-100 hp have been successfully used with laser controlled systems.

Drag Bucket: The leveling bucket can be either 3 point linkage mounted or pulled by the tractor's drawbar. Pull type systems are preferred as it is easier to connect the tractor's hydraulic system to an external hydraulic ram that connects to the internal control system used by the 3 point linkage system. Bucket dimensions and capacity vary according to the available power source and field conditions. A 60 hp tractor has capacity to pull a 2 m wide x 1 m deep bucket in most of the soil types. The design specifications for the bucket should match the available power from the tractor.

Laser Transmitter: The laser transmitter is mounted on a tripod which allows the laser beam to sweep above the tractor unobstructed. With the plane of light above the field, several tractors can work with one transmitter.

Laser Receiver: The laser receiver is an omni-directional receiver that detects the position of the laser reference plane and transmits these signals to the control box. The receiver mounts on a manual or electric mast attached to the drag bucket.

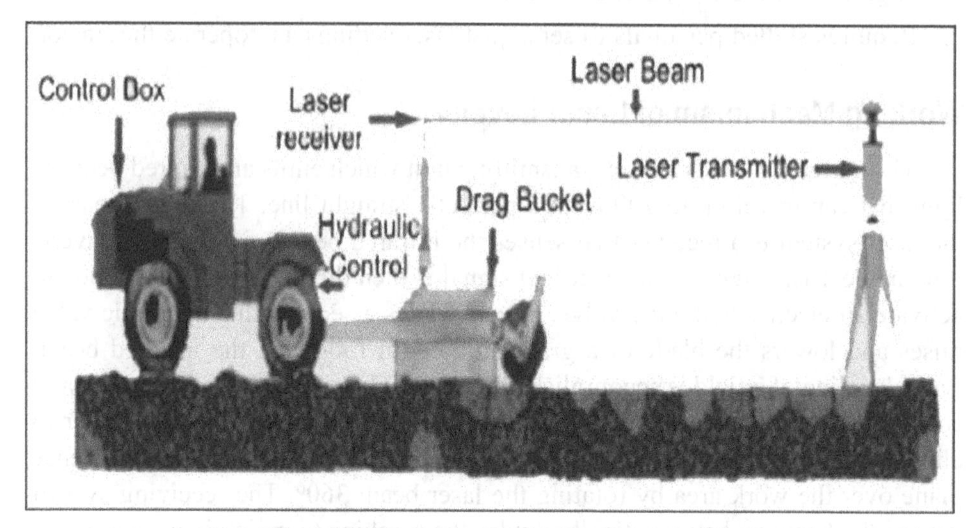

Working Principle of laser land leveling

Control Panel: The control box accepts and processes signals from receiver mounted over the machine. It displays signals to indicate the drag bucket's position relative to the finished grade. When the control box is set to automatic, it provides electrical output for driving the hydraulic valve. The control box is mounted on the tractor within easy reach of the operator. The three control box switches are On/Off, Auto/Manual and Manual Raise/Lower, which allows the operator to manually raise or lower the drag bucket.

Hydraulic Control system: The hydraulic system of the tractor is used to supply oil to raise and lower the leveling bucket. The oil supplied by the tractor's hydraulic pump is normally delivered at 2000-3000 psi pressure. As the hydraulic pump is a positive displacement pump and always pumping more oil than required, a pressure relief valve is needed in the system to return the excess oil to the tractor reservoir. If this relief valve is not large enough or malfunctioned, the damage can be caused to the tractor's hydraulic pump. Wherever possible, it is advisable to use the external remote hydraulic system of the tractor as this system has a built-in relief valve. Where the oil is delivered directly from the pump to the solenoid control valve, an in-line relief valve must be fitted before the control valve. The solenoid control valve when supplied by the laser manufacturers has a built-in relief valve. The solenoid control valve controls the flow of oil to the hydraulic ram, which raises and lowers the bucket. The hydraulic ram can be connected as a single or double acting ram, when connected as a single acting ram, only one oil line is connected to the ram. An air breather is placed in the other connection of the ram to avoid dust contamination on the non-working side of the ram. In this configuration, the weight of the bucket is used for lowering. The desired rate at which the bucket raises and lowers will depend on the operating speed.

Laser Transmitter

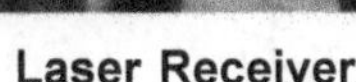

Laser Receiver

Control Panel

Hydraulic Control system

Bucket speed needs to be adjusted with ground speed. The rate at which the bucket will raise and lower is dependent on the amount of oil supplied to the delivery line. Where a remote relief valve is used before the control valve, the pressure setting on this valve will change the raising/ lowering speed. Laser manufacturers usually supply control valves with pressure control adjustments on both the bypass relief valve and the raise and lower valves. When using a hydraulic ram, the ram should be positioned in such a way so that the ram body is connected to push the bucket frame rather than the depth control wheels.

Other equipments/tools may be needed while using a laser system for topographic surveying and these are:

Tape: One 100 meter tape, preferably white metal tapes for more accurate results.

Staff: If a measuring rod is not available. The staff is preferred as metric, upright with an E-type pattern. Check the accuracy of the scale on the staff using a steel tape.

Compass: A compass will be required if direction and bearings are to be recorded. The compass can be used to set magnetic north on the level and allow recordings to be taken from it.

Pegs/Hammer: Pegs, preferably painted white, are required especially for marking out a grid survey or temporary marks.

Book: A notebook is required to record all measurements and other information pertaining to the survey work of the field.

Pencil/Eraser: A pencil and eraser should be preferred over an ink or biro type pen in the field.

Operations Involved in Laser Leveling of Field

A. Ploughing the field
B. Conducting a topographic survey
C. Checking/calibration of a laser transmitter
D. Leveling the field

A. **Ploughing the field:** The field is ploughed preferably from the centre towards outside. It is preferable to plough the field when the soil is at field capacity moist otherwise a significantly more tractor power is required. If the soil is very dry, it is better to do a one way disc harrowing in the field. Disc harrows or tine implements are ideal for second workings. All surface residues should be removed for easy flow of soil through the bucket of laser leveler.

B. **Conducting a topographic survey:** Once the field is ploughed, it is important to conduct a topographic survey to record the high and low spots in the field. After recording the survey readings, one can then establish the mean height of the field by taking the sum of all the readings and dividing by the number of readings taken. Thereafter, with the use of field diagram and the mean height of the field one has to make a plan a strategy for effective movement of soil from high to low areas of the field. Lasers are now widely used to accomplish a topographic survey as they are very accurate, simple to use and readily available in India also. Recordings can be taken up to a radius of 300 meters from the transmitter. The laser surveying system is made up of a laser transmitter, a tripod, a measuring rod and a small laser receiver. A major advantage of laser surveying is the accuracy, simplicity of use and only one person can perform all the operations.

C. **Calibration of a laser transmitter:** The laser transmitter should be periodically checked for the accuracy. Most laser transmitters have two horizontally levelled adjustment screws that allow minor adjustments to be made along the two axes of the horizontal plane. The axes are usually labeled

"X" and "Y". All checking and calibration procedures are done at the zero slope reading.

Items required for checking the accuracy of transmitter:

1. A suitable tripod that allows the rotation of transmitter in 90° increments.
2. A minimum 65 m range should be unobstructed and as close to flat as possible.

Calibration Procedure:

1. Mount the unit on a tripod at one end of the 60 m range and level it. Set "X" and "Y" axes grade counters at zero. With auto leveling transmitters, turn the transmitter control switch to the AUTO position and wait for the Auto Mode Indicator Lamp to stop flashing.
2. Station a rod man with a receiver at the other end of the range 60 m away.
3. Align the laser using the sighting scope or groove, in such a way that the "X" is pointed directly at the rod man. Make sure the pentamirror is rotating and the Auto Mode Indicator Lamp has stopped flashing (if appropriate).
4. Ask the rod man to take a precise reading to within 2 mm and mark the reading as 'X1'.
5. Rotate the transmitter 180° and wait at least 2 minutes for it to re-level. In non-auto leveling transmitters, re-level the transmitter manually. Ask the rod man to take another accurate reading and mark it as 'X2'.

Interpretation:

- If the difference between 'X1' and 'X2' is less than 6 mm, no necessary adjustment is required and the laser can be assumed to give the correct reading.
- If the difference lies between 6 mm and 38 mm, the transmitter then needs calibration and this can be done in the field itself (Follow calibration procedure).
- If the difference is 38 mm or greater, the unit must be re-calibrated at an authorized service centre.

Methodology for using a Laser Leveler:

1. Open the tripod legs and adjust the individual positioning of the legs until the base plate is relatively levelled. Use the horizon as a visual guide to get the base plate level.
2. Attach the laser transmitter to the base plate.
3. If the laser is not self-leveling, adjust the individual screws on the base of the transmitter to get the bubble into the center of both circles. Most lasers will not rotate unless the transmitter is levelled.

4. Once the transmitter is levelled, attach the receiver to the leveling bucket and activate the sound monitor.
5. The laser is now ready to record heights.

D. Leveling of a field involves the following steps:

1. The laser controlled bucket should be positioned at a point that represents the mean height of the field.
2. The cutting blade should be set slightly above ground level (1-2 cm).
3. The tractor should then be driven in a circular direction from the high areas to the lower areas in the field.
4. To maximize working efficiency, as soon as the bucket is nearly filled with soil, the operator must turn and drive towards the lower area. Similarly as soon as the bucket is nearly empty, the tractor should be turned and driven back to the higher areas.
5. When the whole field has been covered in this circular manner, the tractor and bucket should then do a final leveling pass in long runs from the high end of the field to the lower end.
6. The field should then be re-surveyed to make sure that the desired level of precision has been attained.
7. In wet areas where there is poor traction or a chance of bogging the tractor, care must be taken to fill the wet areas from the effected edge in a circular motion.

"TROUBLE SHOOTING"

Problem	Solution
If bucket doesn't rise or lower	· Check the transmitter workability · Check hydraulic connections · Check electric connections on solenoid · Check pressure relief valve setting on control valve· Check for contamination in oil lines
If bucket doesn't respond in certain parts of field	· Line of vision between transmitter and receiver is blocked · Receiver is at the same height as tractor cabin. · Laser beam is above or below the receiver height
If bucket moves in one direction only	· Check hydraulic connections · Check electric connections on solenoid · Check pressure relief valve setting on control valve · Check for contamination in oil lines
If bucket shudders on start at first	· Oil is cold or there is no load in bucket · Check pressure relief valve setting

Laser leveling of agricultural land is a recent resource-conservation technology. It has the potential to change the way food is produced by enhancing resource use efficiency of critical inputs without disturbing and causing harmful effects on the productive resilience of the ecosystem.

In spite of several direct and indirect benefits derived from laser land leveling technology, it is yet to become a popular farming practice in the developing and the underdeveloped countries. For accelerating its popularization and large-scale adoption, it requires a number of well considered and synchronized research, extension, participatory, economic and policy initiatives keeping in view the long term sustainability of our production systems.

Popularization of this technology among farmers in a participatory mode on a comprehensive scale, therefore, needs appropriately focused attention on priority basis along with requisite support from researchers and planners. The change in our vision of future agriculture in relation to food and nutritional security, environmental safety and globalization of markets demands improving resource use efficiency considerably to reach the desired growth levels in food production and agricultural productivity. Laser leveling is evidently one of the ways by which we can address these issues to a great extent.

4

MECHANIZED SOWING: TERMS, TYPES OF MECHANICAL SEEDERS AND THEIR USE

Methods of Planting

Various methods of sowing/planting are practiced in farming systems which can be put under broad classifications such as direct seeding vs. transplanting, direct planting vs. indirect planting and manual vs. mechanized planting.

Direct Seeding vs. Transplanting

In general, a farmer willing to raise a crop has to decide upon selection of appropriate planting methods, which can either be direct seeding or transplanting. Direct seeding refers to as planting of a crop with the use of seeds while transplanting is planting with the use of pre-grown seedlings or plants raised from seeds. Therefore, the two methods can also be described as direct planting and indirect planting. The term direct seeding, also known as direct sowing, refers to the planting of seeds (*e.g.* garden peas) or vegetative planting materials (*e.g.* potato) directly in the soil. For transplanting, it is indirect because the seeds are not immediately and directly sown in the field, instead, these are first set to raise seedlings on raised beds or plug trays outside or inside the greenhouse and then such seedlings are finally planted in the field after getting appropriate size of seedlings. The actual field planting can be accomplished either manually or by mechanical means.

Selection of direct seeding or transplanting for field planting depends on many factors *viz*., the crop species to be grown, ease in planting and survival rate, farmer's familiarity, timeliness, financial capability of the farmer and return on investment.

In addition, the term transplanting is also used to refer to the practice of replanting of an already established plant from one location to another. Transplanting is also convenient with a few plants that can be transferred with a ball of soil around the roots. Heavy duty mechanical transplanters have been invented for large trees.

Direct Seeding Systems

Direct seeding is a relatively new approach in crop production system. Direct seed cropping systems are characterized by seeding with minimum soil disturbance, retaining most of the crop residues in the soil and extended crop rotations. Direct seed farming is defined as any method of planting and fertilizing the crop with minimum or no prior tillage to prepare the soil. There are two systems, first one includes planting and fertilization of crop directly and simultaneously into undisturbed soil in a single operation. Second system undergoes fertilization in soil first and then sowing/planting of crops. Direct seeding system improves farm efficiency by saving time and fuel, reducing labour, extending life of farm machinery and reducing equipment support costs, building soil quality and improving crop yields.

Important Steps for Successful Direct Sowing:

- Ascertain germinability of seeds to obtain the production target.
- Ascertain if seeds can be sown as single unit or will require multiple seeds to reach the production target.
- Use of thoroughly processed seeds and treated seeds (Sometimes to break dormancy other than fungicides/insecticides treatments).
- Sow seeds, ideally centering the seeds in each con-tainer. Some seeds require a specific orientation for optimal growth and development, if so, make sure seeds are sown in the correct orientation.
- Depending on the light requirements of the species, cover seeds with the correct amount of mulch.
- Gently water the seeds with a fine watering head to press them into the growing media.

Selection of Appropriate Mechanical Seeders for Planting

The first mechanical seed drill was developed by Henry Smith in 1850. Today, most field sown crops are seeded mechanically. Selection of a seeder is determined by the following:

1. Size and shape of the seed.
2. Soil characteristics.
3. Total acreage to be planted.
4. Need for precision placement of the seed in the row.

Mechanical seeders contain three basic components: a **seed hopper** for holding seeds and a **metering system** to deliver seeds to the **drill.** A drill opens the furrow for planting of seeds. The drill controls seeding depth and must provide good seed-to-soil contact while minimizing soil compaction that might impede seedling emergence. The most commonly used drill is a simple "Coulter" drill that places seeds into an open furrow. Mechanical seeders are available as either **random or precision seeders**.

Random seeders: Seeders that use gravity and tractor speed to place seeds in the soil are called random seeders. Random seeders meter seed in the row without exact spacing. They are less complex than precision seeders and are useful when spacing between plants in the row is not critical and thinning is not applied to achieve final plant stand as in crops like spinach, coriander, fenugreek *etc*. Random seeders use gravity to drop seeds through holes located at the bottom of the hopper and the size of these holes and tractor speed determine the seeding rate.

Precision seeders: Precision seeders use belts, plates or vacuum to place single seeds at desired seed spacing. These seeders selectively meter seed from the hopper to maintain a preset spacing in the row and can greatly reduce the number of seeds required per unit area compared to random seeding. For example, to achieve the same stand in lettuce, seeding rates can be reduced by 84% using precision seeders compared to random seeders. Precision seeders use a separate power take-off on the tractor drive to power the planter and control seeding rate. Several types of precision seeders like belt, plate, wheel and vacuum seeders are available.

Belt seeder: This type of precision seeder uses a continuously cycling belt that moves under the seed supply. Holes in the belt at specified intervals determine seed spacing. During perfect and correct operation, one seed moves by gravity to occupy one hole on the belt and is released as it passes over the furrow.

Plate seeder: This type of seeder also uses gravity to fill holes in a metal plate rotating horizontally through the seed hopper. The number of holes in the plate and the speed of plate rotation determine seed spacing.

Wheel seeder: This seeder employs a rotating wheel oriented in a vertical position at a right angle to the bottom of the seed hopper. Seed fills the opening at the top of the wheel (bottom of the hopper) by gravity and is carried 180°, where it is deposited into the furrow opening.

Vacuum seeders: These are replacing gravity seeders in the vegetable industry because they can deliver single seed more precisely at a specified row spacing especially in small seeded crops like tomato, irregularly shaped crops like lettuce or uncoated seeds. The vacuum seeder utilizes a vertical rotating plate in the hopper with cells under vacuum that pick up a single seed. Seeds are released into the planting furrow by removing the vacuum in the cell as it rotates above the seed drop tube or planting shoe.

A single row Planter

A row seeder for crops such as corn

A multiple row seeder for small seeds

A precision row seeder for vegetables

Coustesy: https://aggie-horticulture.tamu.edu

A vacuum seeder showing the rotating plate that picks up and delivers single seed.

Coustesy: https://aggie-horticulture.tamu.edu

Presently, manual transplanting method is being adopted at most of vegetable cultivating areas. In manual transplanting, one has to make a dig by hand or a farm tool and place the seedling in the soil and cover it. All operations need to be done in bending posture, which seems to very tedious, and labour and time consuming, it may consume 250 to 350 man hours per ha. Semi automatic vegetable transplanters are now available in India for bare root, plug and pot type of seedlings, however, they are very costly for small farmers. Different planters have been developed to provide a solution to the manual operation at small farms.

Pusa Carrot Planter

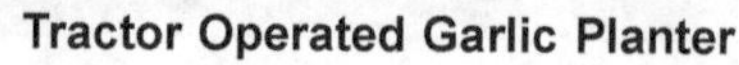

Tractor Operated Garlic Planter

Tractor operated Okra Planter

Pneumatic Precision Planter

Calculating Sowing Rate

The calculation of seed rate is very important factor deciding the cost and precision seeding of calculated amount of seed. The following formula is useful in calculating the rate of seed sowing:

Density desired (plants/unit area)								
Weight of seeds to sow per unit area	=	*Purity Percentage	x	*Germination percentage	x	*Field factor	x	Test Weight/Seed count (number of seeds per unit weight)
*Expressed as a decimal								
Field factor is a correction term that is applied based on the expected losses that experience at that nursery indicates will occur with that species. It is a percentage expressed as a decimal.								

Mechanical Seeding in plug nursery: There has been general practice of sowing seeds either over the surface of flat beds or in rows on raised beds for seedling production. Advantages of row planting are reduced damping-off, better aeration, easier transplanting and less drying out. Planting at very high density as in flat beds encourages damping-off, makes transplanting more difficult, and produces weaker, non-uniform seedlings. Simultaneously seedling raised in soil media generally faces problems of soil-borne diseases particularly nematodes. Therefore, to overcome such problems, there has now been shift towards raising vegetable and flower seedlings/plantlets in plug trays for outdoor as well as indoor cultivation.

The objective of plug tray seedlings production is to get uniformly sized seedlings in each cell. Nurseryman also faces problem in placement of individual seed in each plug of tray. Therefore, mechanical seed sowing in plug trays becomes very important in order to fulfill the demand of farmers at right time. The choice of mechanical seeder depends on several factors like cost, seeding speed, number of plug trays to be seeded and the need for flexibility to sow a variety of seed shapes and sizes. When evaluating a mechanical seeder, nurseryman must consider the machine's ability to deliver seeds at the desired speed without skipping cells due to poor seed pickup or delivery, sowing multiple seeds per cell and sowing seeds without seed bounce that can reduce the precise location of the seed in each plug cell.

Template, needle, and cylinder (drum) are commonly available mechanical seeders for plug nursery raising. Among these, the template seeder is the least expensive type of seeder and it uses a template with holes that match the location of cells in the plug flat. Template seeders use a vacuum to attach seeds to the template. Releasing the vacuum drops the seeds either directly into the plug flat or into a drop tube to precisely locate seeds in each cell of the plug flat. Templates with different size holes are available to handle seeds of different size and shape. A differently sized template is also required for each plug flat size. It is a relatively fast seeder because of sowing an entire plug flat at once. However, this is the least mechanized of the commercially available seeders. It requires the operator to fill the template with seeds, remove the excess and then move the template to the plug flat for sowing. Template seeders work best for round, semi round, or pelleted seeds.

The **needle seeder** is an efficient, fully mechanical and moderately priced seeder requiring little input from the operator. Under vacuum pressure, individual needle or pickup tip lifts single seed from a seed tray and drops one seed directly onto each plug cell or into drop tubes for more accurate seeding. A burst of air can be used to deposit seeds and clean tips of unwanted debris. The needle seeder can sow a variety of seed sizes and shapes including odd shaped seeds like marigold, dahlia and zinnia. Though, needle seeder is slower than the cylinder seeder but still can sow 1,00,000 seeds per hour. Small and medium nurserymen prefer needle seeders because of their cost and flexibility in seeding.

Sowing of Seeds on a Tray Flat in Single Operation by Template Seeders

Vacuum removal of a line of seeds from a tray and dropping onto tray flat by needle seeders. Direct Seed placement onto each cell or indirect placement via drop tubes

Drum seeder: Vacuum picking and dropping of a line of seeds onto the plug tray by a large rotating drum

Coustesy: https://aggie-horticulture.tamu.edu

Precise adjustment of seed placement by computer software as per the size of plug tray

Vacuum tubes in cylinder type of seeder

Adjustment of sowing rate and placement without removing the cylinder

A seeder set to sow two seeds per cell

Cylinder/drum seeder

Coustesy: https://aggie-horticulture.tamu.edu

The **cylinder** or **drum seeder**s have a rotating cylinder or drum that picks up seeds from a seed tray under vacuum and drops one seed per plug of the tray. These are fully mechanical, fastest, highly precise but the most costly than other seeders. Most drum seeders require a different drum for each plug flat, but newer models of cylinder seeders have several hole sizes per cylinder that can selectively be put under vacuum pressure and can be adjusted for different flat types via computer. These can sow single or multiple seeds per plug cell at a time depending upon the requirement. Sophisticated seeders eject seeds from the drum using an air or water stream for precise seeding location in the flat. These seeders work best with round, semi-round, or pelleted seeds. Large nurserymen have the capacity to sow millions of seeds of more than 100 varieties of crops in plugs per year, so they must prefer cylinder seeders because of their ability to sow seeds up to 800,000 seeds per hour very quickly. Cylinder seeders are becoming more common than drum seeders because they offer more flexibility

IIHR, Bengaluru: New Developments

Dibbler cum seeder for plug trays

Automatic plug tray filling, dibbling, seeding and watering machine

Courtesy: https://www.iihr.res.in

5

APPROACHES FOR MAPPING OF SOIL AND PLANT ATTRIBUTES

SOIL MAPPING

Soil map refers to a geographical illustration of multiplicity of soil types and soil properties like soil texture, EC, pH, organic matter, depths of horizons *etc*. Generally, soil survey serves as a prerequisite for the formulation of effective soil maps. Soil maps are mostly used for land evaluation, spatial planning, agricultural extension, environmental protection and other similar projects. Traditional soil maps are generally equipped to show general distribution of soils only with soil survey report. Many soil maps are now being derived using digital soil mapping providing richness in context and showing higher spatial detail than the traditional soil maps.

Soil maps are now available in various digital formats, which have various applications in geosciences and environmental sciences and their use can also be extended to agriculture. In modern applications, soil maps are stored as soil resource inventories in Soil Information System (SIS) majorly containing soil geographical database. A SIS is a systematic collection of values of the targeted soil variables available for the entire area of interest and supported with report, user manual, metadata *etc*. In most of the cases, it is a combination of polygon and point maps linked with attribute tables for profile observations, soil mapping units and soil classes. There is also a provision of manipulation for different components of a SIS, which enables better visualization of grids or polygons (spatial references). In the case of pedometric mapping, values of geographic location, soil depth *etc*. can be simulated and envisaged at 2 D or 3 D levels for GIS modeling.

Objectives of Soil Maps

1. Help to produce the key properties of the soils in a single document using a set of codes and legends for quick interpretation of the results

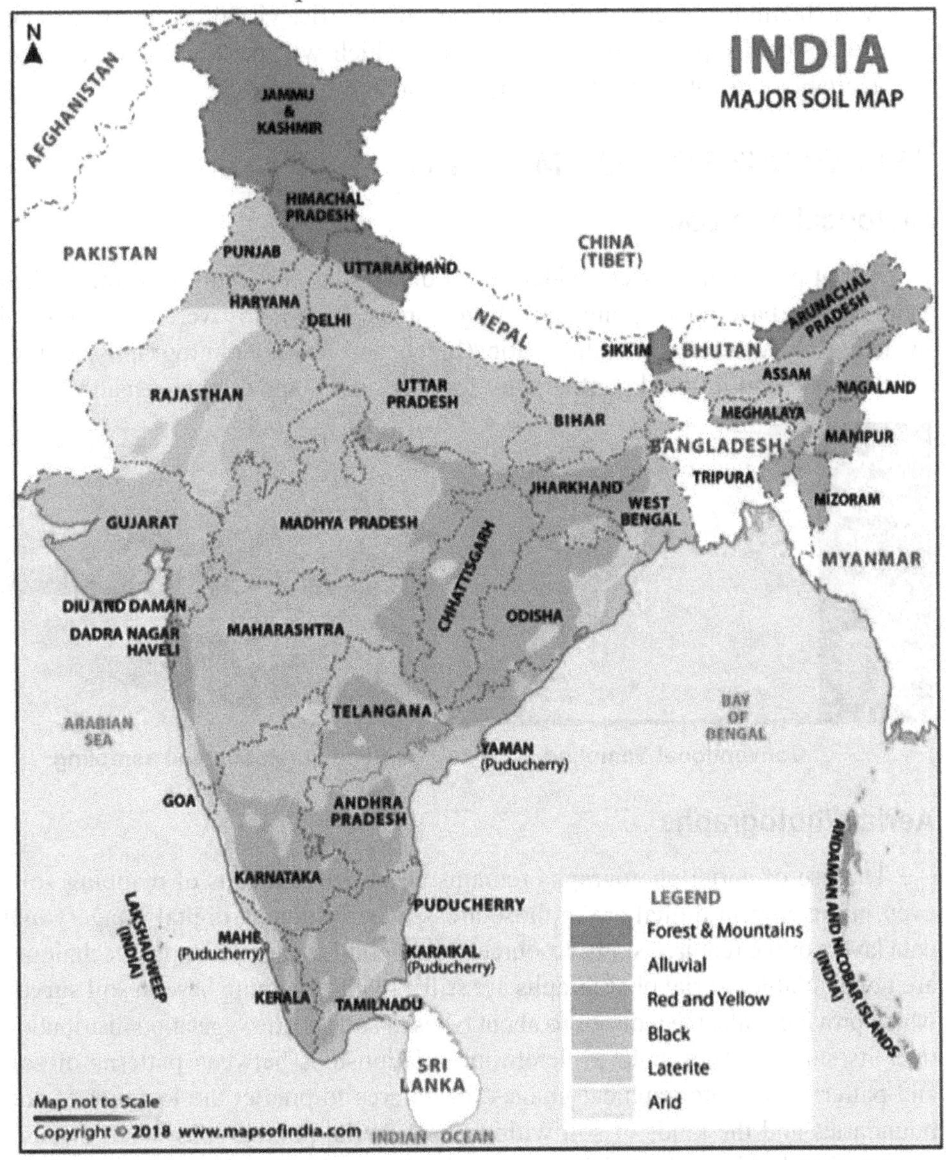

Courtesy: www.mapsofindia.com

2. Describe the distribution of soil units by making groups of more or less generalized or detailed concepts with a suitable working scale
3. Provide a graphic manuscript for preparing the inventory or evaluation of the soil units with their corresponding distribution over the land
4. Help in dissemination of knowledge regarding spatial distribution and properties, which can rapidly be accessed by a non-specialist

5. As it facilitates access to information on specific, visual, synthetic and bi-dimensional representation of the soils, which will permit coherent use of available soil data for better interpretation and results.

APPROACHES OF SOIL MAPPING

Historical Approach

Aerial photographs were widely used during the 20th century in the United States as the base for mapping soil survey areas. Base maps were also prepared by using conventional panchromatic (black and white) photography, colour photography and infrared photography for soil survey and remote sensing.

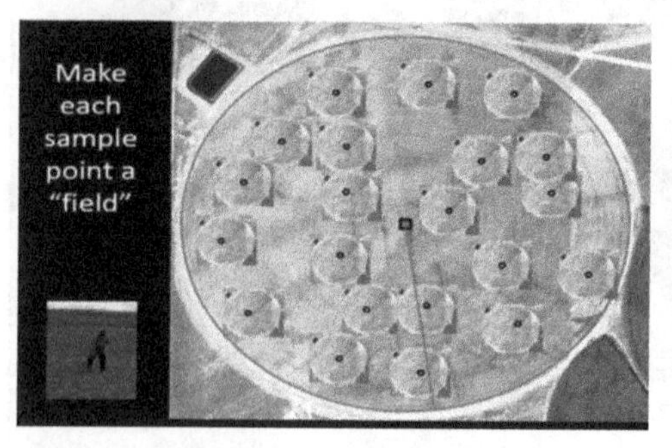

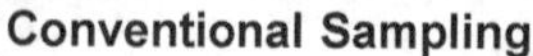
Conventional Sampling

Semi-automatic soil sampling

Aerial Photographs

The use of aerial photographs remains an effective means of mapping soils even in the current digital era in those areas, where suitable digital imagery and data layers or the required skills, resources or support for digital mapping techniques are not available. Aerial photographs are still a viable mapping base in soil survey which provides valuable indications about types of soil and the vegetation distribution over its surface. They help to determine relationships between patterns of soil and patterns of photographical images for an area to predict the location of soil boundaries and the kinds of soil within them. Aerial photographs taken through spectral bands like coloured infrared make it possible to observe subtle-slight variations in the plant populations. Other infrared spectral bands are helpful in characterizing diversity in mineralogy and moisture in the soil. The relationships of these visual patterns on the photographs can be interpreted corresponding to the soil characteristics found by inspection at ground level.

Aerial photographs recognize roads, buildings, lakes, rivers, field boundaries and many kinds of vegetation of a particular location. Cultural features commonly are the easiest features to recognize on aerial photos, but they generally do not coincide precisely with differences in soils, except in areas with significant anthropogenic alterations. Accurate soil maps can't be produced merely by interpretation of aerial photographs because time and place influence the indications visible on such photographs. It is well known fact that human activities have changed the patterns of vegetation and baffled their relations to soil patterns. Therefore, photographical indications must be correlated with soil attributes and verified at field level.

Contemporary Approach

Photographs have now been swapped by digital imagery in the 21st century as the mapping base in soil survey. The digital imagery has ability to super impose multiple imagery resources for comparative analysis with rapid adjustments. The raster-based soil maps have now magnified the speed of delivering soil survey products. The choice of background imagery like colour and topographic imagery have also improved and customized the soil survey products to display improved information.

Digital soil mapping generates geographically referenced soil databases on the basis of quantitative relationships between spatially unambiguous environmental data and the measurements made in the field and laboratory. The digital soil map is a database (raster) composed of 2 D cells (pixels) organized into a grid depicting a specific geographic location in each pixel with supporting soil data. Digital soil maps illustrate the spatial distribution of different soil classes or properties and have ability to predict the uncertainty of the soil at field level. Digital soil mapping can be used to create initial soil survey maps, refine or update existing soil surveys, generate specific soil interpretations and assess risk. It can facilitate the rapid inventory, re-inventory and project based management of lands under changing environment conditions.

Stages of Digital Soil Mapping

Each stage of digital soil mapping consists of defined objectives, which requires review and assessment at several points for accomplishing the whole process of soil mapping. The well defined processes of each stage are required to be followed for making any digital soil mapping project a success and these are:

1. Defining the area and scope of the project
2. Identifying physical features of interest
3. Identifying sources of data and its pre-processing
4. Exploring the data and conducting landform analysis

5. Sampling for training data
6. Predicting classes or properties of soil
7. Statistical analysis for checking the accuracy of data
8. Application of digital soil mapping

Digital vs. Conventional Soil Mapping

The availability and accessibility of geographic information systems (GIS), global positioning systems (GPS), remotely sensed spectral data, topographic data derived from digital elevation models (DEMs), predictive or inference models and software for data analysis have potentially improved the art and science of soil survey. Conventional soil mapping uses point observations in the field that are geo-referenced with GPS and digital elevation models visualized in a GIS. The important distinction between digital soil mapping and conventional soil mapping is that digital soil mapping uses quantitative inference models to generate predictions for different soil classes or properties in a geographic raster. Models based on data mining, statistical analysis and machine learning organize vast amounts of geospatial data into meaningful clusters for recognizing spatial patterns.

Limitations of Conventional Soil Mapping

- Soil-landscape model generated and knowledge are implicit, intuitive and not documented properly, hence has the limitation to re-build and reproduce.
- Soil class/type or cloropletic (polygon) maps produced through conventional soil mapping try to give prediction for representative sample (modal profile).
- Assumptions used under conventional soil mapping have limitations of unknown accuracy.
- Variability inside a soil class is considered non-spatially correlated.
- There is lack of quantitative expression of the spatial variability of soil and its properties.
- Two main types of errors occur in conventional mapping and these are commission and omission errors.

Commission errors: Establishing a Minimum Legible Area (MLA) and eliminating smaller areas.

Omission errors: Eliminating spatial variability of soil properties inside a polygon in reference to GIS.

Terrain attributes are derived from DEMs and are typically represented using the raster data format. Elevation can also be represented as points [*e.g.*, LiDAR

(Light Detection and Ranging) returns] or triangulated irregular networks (TIN), but the raster format is typically preferred due to its greater flexibility. Elevation data are typically developed from contours, topographic surveys, or LiDAR data. Terrain attributes may be broadly grouped into two categories: (1) primary attributes, which are computed directly from a DEM; and (2) compound attributes, which are combinations of primary attributes. There are some primary and compound terrain attributes used in digital soil mapping very much relevant to depict measurements with biophysical properties.

Applications of Digital Soil Mapping

Digital soil mapping is extensively used to produce soil maps predicting soil classes and properties. However, the process of generating spatially explicit predictions of natural phenomena using quantitative relationships between training data and predictor variables can be applied to create a broad spectrum of information products. The main application of digital soil mapping in pedology and related fields lies in producing information products other than soil maps. The maps so generated may be used in one or other way to depict their applications:

Raster vs. Polygon, Disaggregation and Evaluation of Existing Maps

It is very important in relation to updating soil surveys, which helps to generate predictive modeling maps.

Predicting Biological Soil Crusts

Biological soil crusts are communities of cyanobacteria, algae, micro-fungi, mosses, liverworts and lichens at the soil surface. They stabilize soil, minimize wind and water erosion and are important sources of N and organic C in arid and semi-arid ecosystems. So, digital mapping models help to predict these important communities of a particular biological crust for better use of such organisms in agricultural applications.

Predicting Ecological Sites

Correlating ecological sites with soil map units is an important component of soil mapping as it provides better understanding on the interaction of biotic factors with abiotic ones in the environment. This could be very useful in many land management decisions.

Selected Primary and Compound Terrain Attributes Used in Digital Soil Mapping

Attributes	Measures	Biophysical properties
Primary:		
Curvature	Second derivative of slope	Flow characterization, i.e., runoff or run-on
Relief/Topographic Ruggedness	ABS (Zmax-Zmin) for specified neighbourhood	Broad characterization of terrain (infers parent material)
Normalized Slope Height/Relative Elevation or Relative Position	(Z- Zmin)/ (Zmax- Zmin) where Z is the elevation of central cell for specified neighbourhood	Relative landform position, catenary sequence, vegetation distribution
Compound:		
Solar Radiation	Estimates potential or actual incoming solar radiation for specified time interval	Solar energy incidence on surface, a means of modelling aspect
Wetness Index or Topographic Wetness Index	W = (A/S); where A is upslope contributing area for a cell and S is the tangent of slope gradient	Spatial distribution of zones of saturation for runoff (assumes uniform soil transmissivity within the catchment)
Potential Drainage Density	Cell count of stream segments within specified neighbourhood	A measure of landscape dissection
Morphometric Protection Index	A measure of topographic openness	Plant communities, soil development, impact of wind
Multi-Resolution Valley Bottom Flatness Index and Ridge Top Flatness Index	Process to differentiate valley floor and ridge top positions	Landscape position
Geomorphon	Landform classification based on line-of-sight	Crisp landform classes, catenary sequence

Source: https://www.nrcs.usda.gov

Predicting Rare Plant Habitat

Digital mapping can be used to generate models to identify potential habitats across a large area specifically in remote terrains, where accessibility is time and labour intensive. Once soil and site data are located for potential habitat areas, they can be used to verify habitat suitability and the conservation or restoration efforts can be initiated for such habitats.

Advantages of Digital Soil Mapping

1. An accurate model can be generated via a process of iterative development and testing to create the final soil map fitting to accuracy and uncertainty standards.
2. The uniform application of the model across area results in a consistent soil map.
3. The accuracy and uncertainty scales of model associated with the soil map can be quantified.
4. Offers collection of information for each grid cell/management zone of selected area, hence reveals more detailed database of narrow soil variability over the area for better management.
5. The models through digital mapping to predict soil classes or properties are effective ways to collect and preserve expert knowledge about soil and landscape relationships in future applications.

MAPPING OF PLANT ATTRIBUTES

Remote Sensing Techniques

The application of remote sensing data started during early 80s in precision agriculture in order to analyze variations in crop and soil conditions. Remote sensing technology has extensively been used during last 3-4 decades for timely assessment of changes in growth and development of various agricultural crops.

Multispectral remote sensing has a significant role in precision agriculture as it represents crop growth condition on a spatial and temporal scale and is also cost effective. It plays significant role in generating and understanding the relationships of crop data like vegetation development, photosynthetic activity, biomass accumulation, leaf area index (LAI) with crop evapotranspiration (ET) for making prediction on crop production. Airborne multispectral remote sensing has been used in assessing the crop yield conditions.

Many pragmatic relationships and equations have been developed between vegetation indices and leaf area index, fractional ground cover and crop growth rates in the past through ground sampling. These relationships are then used by the crop growers to estimate the expected yield of crops prior to harvest to execute crop management decisions for maximizing productivity and market gains. Crop production and its estimation have a direct impact on the food management and economic development of a nation.

Satellite Imagery

High resolution multispectral satellites have now enabled the application and use of satellite data in agriculture sector for crop mapping. Satellite remote sensing covers large areas and the data analysis can be done in a single image thereby consuming less time. Simultaneously, the data can be recorded in different wavebands which provide accurate information about the ground conditions. Though, imagery is available from satellite systems, but there are some distinct drawbacks such as higher cost for smaller spatial extent as well as lower spatial and temporal resolution. Satellite images do have other problems like data masking due to cloud presence, non-availability of data for real time management of crop growth due to fixed temporal frequency and correction of radiometric data owing to atmospheric interference. The normalized difference vegetation index (NDVI) derived from the visible and near infrared (NIR) bands of the NOAA AVHRR satellite has been successfully used to monitor vegetation changes at regional scales. Temporal changes in the NDVI are related to net primary production. The studies have also generated strong equations to relate primary production estimates on the basis of absorption of photosynthetically active radiation (PAR) by the vegetation. Remote sensing is very effective method for non-destructive monitoring of plant growth and detecting many environmental stresses which have direct impact on crop productivity.

Aerial Platform and Unmanned Aerial Vehicle Technology (Drone Technology)

The use of Unmanned Aerial Vehicles (UAVs) has spread very fast for agricultural applications in recent past. Van Blyenburgh has defined UAVs as uninhabited, reusable, motorized aerial vehicles which rely on microprocessors allowing autonomous flight nearly without human intervention.

Modern imaging technology based on UAVs presents unparalleled potential for measuring and monitoring our surroundings. For many applications, UAV based airborne methods are very much suitable for many applications including agriculture because of cost effectiveness of the technology in data collection with desired level of spatial and temporal resolutions. This technology has an important advantage of collecting remote sensing data even under poor imaging conditions (*e.g.* cloud cover/overcast condition), which makes UAV a truly versatile application in a wide range of environmental conditions.

UAVs have been known to deliver high spatial and temporal resolution data required in crop growth monitoring that helps to facilitate detection and quantification of variability within a field to outline agricultural management strategies.

In comparison to satellite remote sensing, aerial imagery is more applicable to precision farming due to the following advantages:

- Helps to acquire images frequently over the cropping area throughout growing season.
- The collection of data is independent of clouds and provides fast data acquisition with real time capability. Also has an option to reschedule image acquisition even to a cloud free day if there is data mask due to heavy clouds on the day of acquisition.
- Offers high spatial resolution expressing soil and crop growth variability for better assessment.
- Its cost per unit area is relatively low when scanning large areas.

Benefits of UAVs outdo its disadvantages in the following way:

- UAVs when equipped with high precision cameras, can help to recognize the reality of a field using a multispectral sensor.
- With their ability to cover distances quickly, drones can reduce the time from days to hours.
- Based on weather trends and predictions, drones can also be proactively positioned in areas of high claim activities.
- Could be very handy in streamlining and settling claims pertaining to crop insurance process as drones communicate information in real time.

Example

Understanding Corn Phenology

The analysis of crop growth during different phenophases is an important component of precision farming. So, the knowledge of various developmental processes of a corn plant is very useful in evaluating its yield potential. Remote sensing has great potential of generating data for such kind of investigations in the field of precision agriculture.

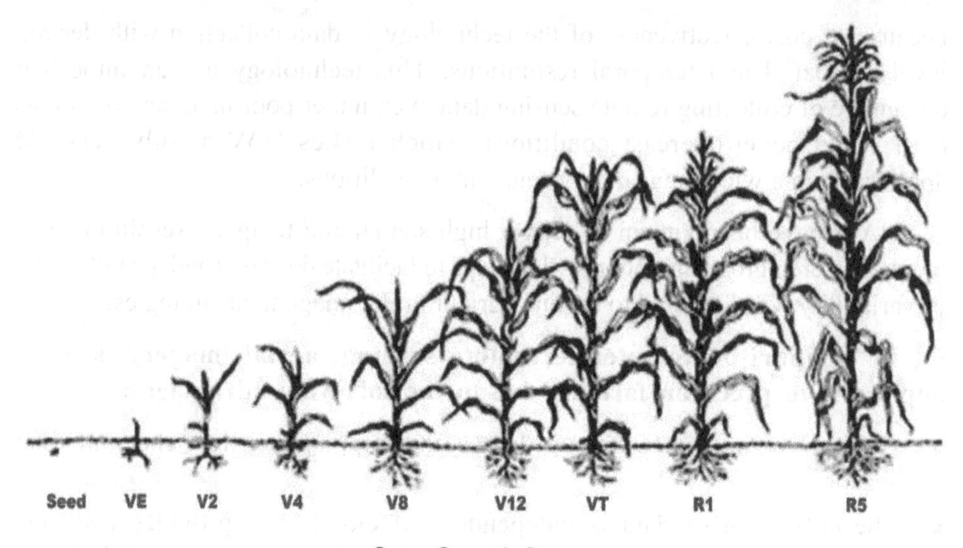

Corn Growth Stages

Vegetative Stages		Reproductive Stages	
Stage	**Description**	**Stage**	**Description**
VE	Emergence	R1	Silking – visibility of silks outside the husks
V1	One leaf with collar visibility	R2	Blister - kernels are white and resemble a blister in shape
V2	Two leaves with collar visibility	R3	Milk - kernels are yellow on the outside with a milky inner fluid
V(n)	(n) leaves with collar visibility	R4	Dough - milky inner fluid thickens to a pasty consistency
VT	Last branch of tassel is completely visible	R5	Dent - nearly all kernels are denting
		R6	Physiological maturity - the black abscission layer has formed

Use of Vegetation Indices (VIs)

Remotely sensed spectral vegetation indices are widely used for the assessment of biomass, water use, plant stress, plant health and crop production. However, the successful use of these indices requires knowledge of the units of inputs used to form the indices and an understanding of relationships of external environment and vegetation canopy with the computed index values. Healthy crops are generally characterized by strong absorption of red energy and strong reflectance of NIR energy. The absorption and scattering of the red and near-infrared bands can be combined into different quantitative indices of vegetation conditions, which are termed as vegetation indices. Since the late 1980s, numerous studies have been conducted on crop growth analysis using normalized difference vegetation index (NDVI) to support precision agriculture.

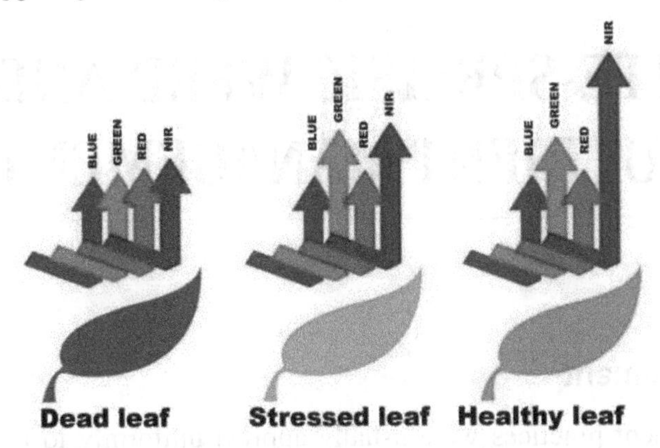

Reflectance Properties of Leaves at different stages

VIs are designed to provide a measurement of the overall amount and quality of photosynthetic material in vegetation essential for understanding the state of vegetation for any purpose. These VIs are can be correlated with the fractional absorption of photosynthetically active radiation (fAPAR) in plant canopies and vegetated pixels. They do not provide quantitative information on any one biological or environmental factor contributing to the fAPAR, but broad correlations have been found between the broadband greenness VIs and canopy LAI.

6

SITE-SPECIFIC WEED AND NUTRIENT MANAGEMENT

Weed Management

Weed management practices were usually applied uniformly to the entire field just like the other management practices adopted for crop, soil and pest. It has been well recognized by the scientists and growers that the presence of weeds is rather patchy *i.e.* not uniform across the fields. So, uniform application of weedicides/ herbicides across the field where target weeds are not uniformly distributed can waste resources and add to the social, environmental and economic concerns about their use. Precision farming proposes so many effective set of tools, which can address these apprehensions and increase the efficiency of weed management. Different methods have been developed to scout and detect weeds so as to adopt control measures at the place of requirement. This can ensure application of weedicides at lower rates with lesser environmental risk and possibly higher social acceptance of farming communities.

Weed Biology and Precision Farming

Weed biology is the study of the life cycle of weeds and the emergence of weeds from seeds or shoots from underground organs such as tubers that are considered to be important facets of weed biology for effective control of weeds. Weeds are known to cause heavy economic losses and the methods usually employed to control severe environmental impacts. Generally, weeds appear in patches because of their differential spread, survival and reproduction within a field over a period of time and these weed patches have tendency to remain in the same place even over the years with probable range of weed density. Precision farming has the potential of targeting the patchy nature of weeds at specific sites without wasting expensive and potentially harmful chemicals or inputs.

Weed Response to Site-Specific Management

The knowledge on variable weed emergence patterns provides basics for site-specific management of weeds. Understanding about various causes of differential emergence of weeds provides an important decision making support for timely operations and precise management. One of the most important features of weeds is non-consistent nature. Growers should always anticipate about the appearance of weeds that may increase at a site because of their management practices. Even after the successful mapping, monitoring and management of weeds, shifts in weed species are likely to occur due so many factors like environmental, cultural *etc.* Over the period of time and depending upon the dispersal, weeds propagules can travel long distance and invade new areas (*e.g.* congress grass). Therefore, field mapping of weeds plays very important role to determine current problems of a place as well as to anticipate the severity of weeds in time to come.

Precision agriculture is a powerful technology, which offers potentially very important and powerful applications for weed management. Utmost care should be taken to adopt not only the good agricultural practices but efforts should also be directed towards the prevention of weed infestations and wise selection of herbicides to check resistance and curtail environmental issues associated with weed management.

"*WeedCast*" Software

Weedcast software constitutes an important part of precision weed management which gives an idea about the timing of weeds germination and the speed of growth after germination. It precisely takes into account the day to day weather data for predicting weed emergence in real time. It combines information on simple weather parameters like daily minimum and maximum air temperatures and rainfall to make such predictions. The peculiarity of the *WeedCast* software is that it takes into account the air temperatures and rainfall values for assessing soil temperature and water content of the upper 5 to 10cm layer of soil, which actually determine time and speed of weed emergence. Nonetheless, it must be connected directly, automatically to interpolated and geo-referenced soil and weather data on daily basis mining the interpretation from the system by the users. Once all the data are available in this software, one can get full information on different weed species likely to appear in a particular environment and this information may help to strategize the planning for effective management of weed species.

Variable Rate Technology for Weed Control

With the advancement of technology, different agricultural industries have are now manufacturing sprayers with specific controllers to minimize the variations of doses of chemicals applied in the fields. The control systems in such modern devices have great control to adjust the changes as per the vehicle speed and regulate variable rates of chemical applications in accordance to the preplanned maps.

Most of the precision application devices used for planned application rates is equipped with GPS receivers to decide the appropriate level of application rates for weedicides for a given area in the field. Two other important components are required to carry out variable rate application of herbicides; first one is 'Task Computer', which is required to give signals signifying the target application rate for the current location and second system is meant to change the application rate physically to counterpart the target rate. There are different control systems which can be used for variable rate application of herbicides as per the details given below:

Flow-Based Control System

The flow based control is the simplest system consisting of a tank connected with a flow meter, a ground speed sensor and a controllable valve (servo valve) along with an electronic controller to manage the application of desired rate of chemical from a tank mix. A microprocessor in the sprayer consists of console and controller which uses the information on sprayer width and desired amount per unit area to work out the suitable flow rate for the current ground speed. The servo valve is then opened or closed until the flow meter measurement matches the calculated flow rate of chemical. A communication link must be recognized between this controller and the map system to establish a variable rate application of chemicals. An illustration of the components comprising such a system is shown in **Fig. 6.1**.

Direct chemical injection system

In this system, application of chemical is done by injecting it into a stream of water. This system comprises of a controller and a chemical pump to manage the rate of chemical injection rather than the flow rate of a tank mix. The flow rate of the carrier *i.e.* water is usually constant and the injection rate of chemical is made to vary according to the ground speed or changes in the directed rate. The controller in this system should be designed in such a way to accept external commands to execute variable rate application of chemicals. This system has the advantage of eliminating leftover tank mix and reducing the risk to chemical exposure. This system has an added benefit of adjusting the constant flow of carrier to nozzles for uniform spray of chemicals with a desirable size of droplets. Direct chemical injection system is depicted in **Fig. 6.2**.

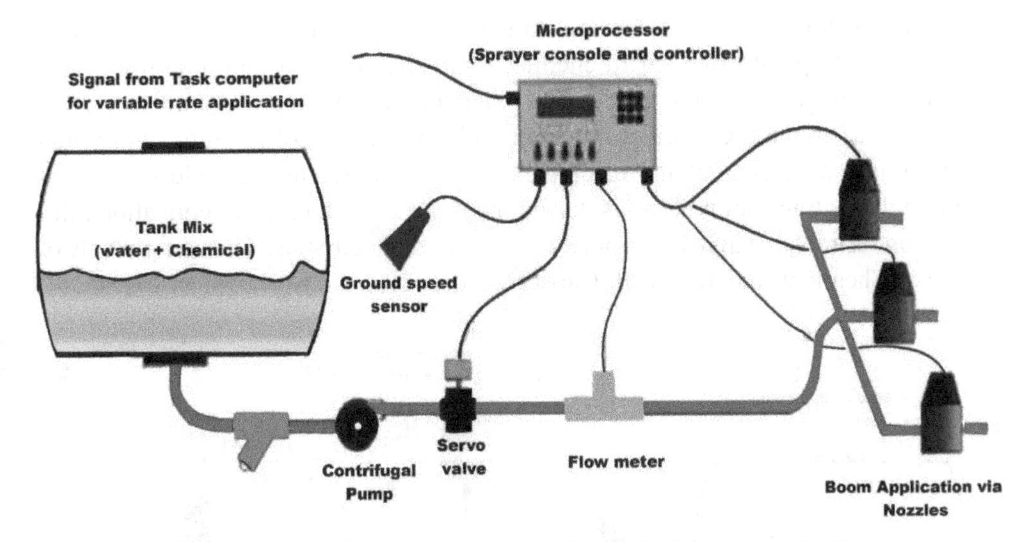

Fig. 6.1: Flow-Based Control System for variable rate application

Courtesy: www.ipni.net/publication/ssmg.nsf

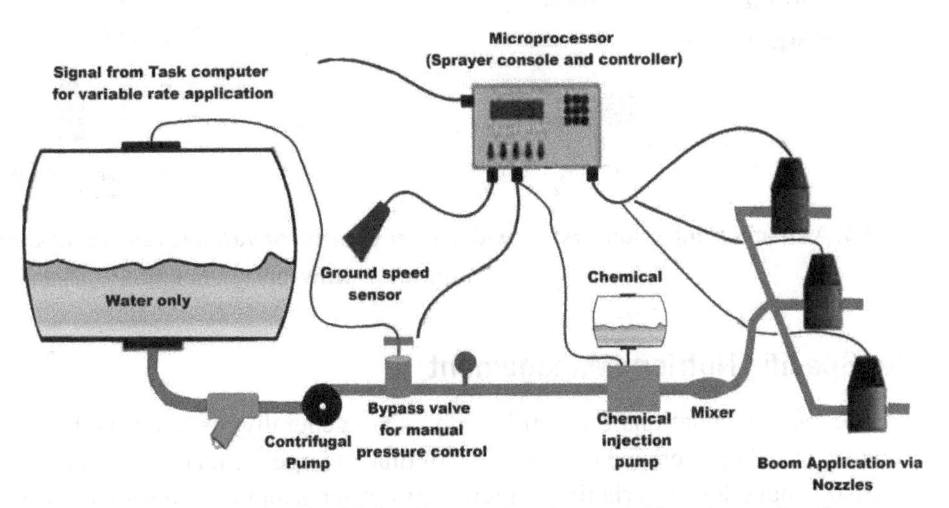

Fig. 6.2: Direct chemical injection system for variable rate application

Courtesy: www.ipni.net/publication/ssmg.nsf

Direct Chemical Injection with Carrier Control

Chemical injection with carrier control has ability to make changes in rates of chemical injection and water in response to the speed or change in application rate. There are two controllers; the first one controls the injection pump while the second controller manages a servo valve in order to provide a matching flow of water. This system delivers a mix of constant concentration of chemical assumed to be coming from a premixed tank. Direct injection of chemical through the system has no issue of leftover mix to worry about and helps to avoid the direct exposure of operator/worker to chemicals in the process of tank mixing. In this system, both chemical and carrier controllers make very fast adjustments to from one rate to another. Chemical injection with carrier control requires high initial investment and has much more complex system in delivering variable amounts of liquid through the spray nozzles during the process of change in application rates with undesirable changes in droplet and spray characteristics. The components of Direct Chemical Injection with Carrier Control system are shown in **Fig. 6.3**.

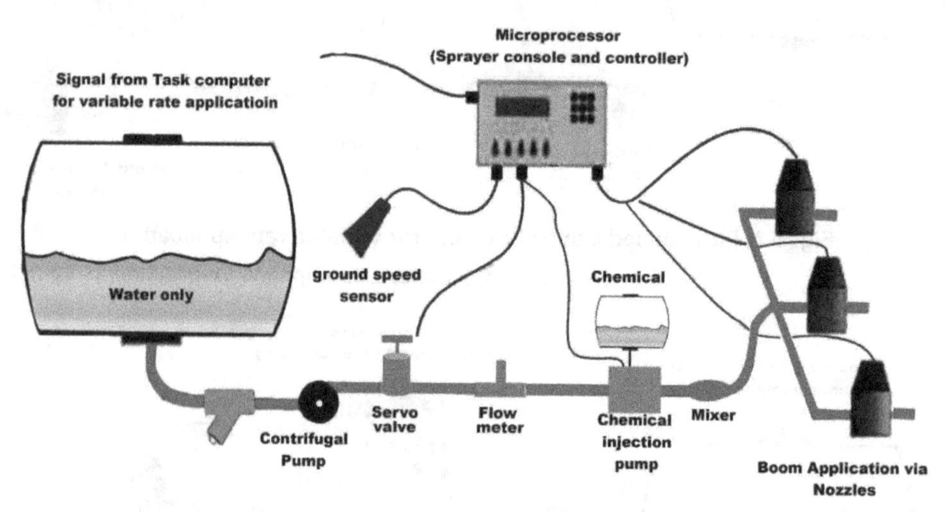

Fig. 6.3: A chemical injection system with carrier control for variable rate application

Courtesy: www.ipni.net/publication/ssmg.nsf

Site-Specific Nutrient Management

The recommendations for fertilizer doses are generally made on the basis of crop response data averaged over a period of time in large area. However, farmers' fields/soils have lot of variability in terms of nutrients and crop response, which actually make farmers to over fertilize in some areas and/ or under-fertilize in others. Site-Specific Nutrient Management (SSNM) provides appropriate guidelines well suited to the reference of farmers' fields, which helps to maintain or enhance yields of crops and concurrently saves inputs cost through more efficient fertilizer use. The adoption of SSNM can avoid injudicious use of fertilizers, which can be very helpful in reducing greenhouse gas emissions to a greater extent even up to 50% in some cases. The main objective of Site-Specific Nutrient Management (SSNM) is to supply optimal nourishment to crop plants as per the actual requirement over time and space through four key principles, designated as "4 R'.

"Key scientific principles and associated practices of 4R nutrient stewardship"

SSNM principle	Scientific basis	Associated practices
Right Product	Application of balanced nutrition suiting to soil properties	Use of commercial fertilizers, farmyard manure, compost and crop residues
Right Rate	Assessing crop demand and supplying nutrients from all sources accordingly	Soil test and or/ plant tissue analysis for nutrients and balancing the application rate with that removed by crop for actual gain in yield
Right Time	Assessing the dynamics of uptake of nutrients by crop species and supply from soil. Determining the timing of loss risk	Application of nutrients appropriately at pre-planting, at planting, at flowering and at fruiting for balancing the growth and development in each important growing stage of crop
Right Place	Recognizing crop rooting patterns Managing spatial variability	Application of nutrients could be through different practices like broadcasting, band or drill application or direct injection, variable rate application

Advantages

1. Site-Specific Nutrient Management (SSNM) optimizes the nutrients supply to match crop requirements over space and time.
2. SSNM helps to improve fertilizers use efficiency thereby increases crop productivity.
3. SSNM helps to mitigate greenhouse gases from areas using high doses of nitrogenous fertilizers.

TOOLS FOR IMPLEMENTING SSNM ON THE FARM

Optical Sensors

Optical sensors are used to get an idea of vegetative index called NDVI (Normalized Difference Vegetation Index) by measuring the reflectance from leaves of a crop. This index interprets the status of nutrients in plants based on their size and color (green versus yellow). Originally, this technology was developed for large farms, however handheld smaller version of such optical sensors are also available for small scale applications. These small hand-held optical sensors can be used by the farmers and extension workers to develop SSNM recommendations particularly for N.

Farmer using a handheld sensor to measure NDVI

Software for SSNM: Nutrient Expert® and Crop Manager

Computer or mobile phone based tools are now increasingly being used to facilitate improvement in nutrient management at farmers' fields, especially in the areas where fertilizer recommendations based on large area are available. Nutrient Expert® and Crop Manager are the examples of decision-support systems developed for SSNM.

Nutrient Expert® is an interactive, computer-based decision support tool that allows smallholder farmers to implement SSNM rapidly in their fields with or without soil test data. This software works on the basis of estimating yield for a farmer's field based on the growing conditions and determining the nutrient balance in the cropping system based on yield and fertilizers/manures application made in the previous crop. So, it combines such informations with nutrient response for expected N, P and K in targeted fields to establish location-specific nutrient recommendations. The software also has the facility of carrying out economic analysis by comparing costs and benefits between farmers' practice and recommended practices.

Crop Manager

Crop Manager is a computer and mobile phone-based application that make site and season-specific fertilizer recommendations to small scale farmers. The tool provides information to the farmers for making adjustments to crop requirements based on soil characteristics, water management and crop variety. Recommendations are based on user-input information about farm location and management, which can be collected by extension workers, crop advisors and service providers.

Challenges of SSNM

Technology and knowledge: It is prerequisite to have thorough knowledge of soil properties and nutrient status of crops, which actually decides the quantity of fertilizer accordingly in SSNM. Therefore, it is mandatory to develop decision support systems, farmer-friendly tools and techniques for using the important information to calculate nutrient requirements so as to make SSNM more accessible to the farmers and/ or the farm managers.

Fertilizers Availability: Accessibility of farmers to different synthetic or organic fertilizers is not easy all the time and the cost of these inputs also varies from one geographical region to other. So, establishment of organized markets for such inputs or identification of on-farm nutrient sources is essential to make the adoption of SSNM more effective. Moreover, SSNM helps farmers to make best use of limited nutrient resources.

Economic uncertainty: SSNM must ensure good savings from reduced fertilizer use without compromising with the yield, in other words the value of increased production should be higher than the costs of adopting SSNM technology. Farmers are more likely interested to get good net returns with high valued crops, where increase in yield can considerably increase profits even during the time of high price for fertilizer inputs.

7

PRECISION MANAGEMENT OF INSECT-PESTS AND DISEASES

Precision Pest Management

Precision Pest Management is defined as an art and science of employing advanced technologies to restrict pest population below the threshold level in order to achieve higher crop yield along with minimum impact on the environment. It aims at managing the existing variability of insect-pests and diseases with judicious use of pesticides at micro-level. The spatial and temporal variability of pest populations are assessed substantially for making decision on their management with utmost care for ecological safety at considerably low costs. As the principle of precision farming specifically emphasizes on 3 R (Right input, right time and right location), so the in-field variability should be managed by applying appropriate dose of pesticides in right amounts, at right time and locations after recognizing, locating, recording and quantifying the variability.

Components of Precision Pest Management

The important components of precision pest management are Geographical information system (GIS), Global Positioning system (GPS), Remote sensing and the farmers/ Producers.

Geographical information system (GIS): GIS is an important tool helpful in deriving values from information generated on pest population dynamics for better interpretation. GIS carries out the spatial analysis of variability thus allowing decision making process easy.

Global Positioning system (GPS): It provides accurate positional information which is useful in locating the existing spatial variability over the field. Thus, targeted application of management techniques can efficiently be made to get good results comparatively at lower cost.

GIS GPS

Remote sensing Farmer

Components of Precision Pest Management

Remote sensing: Remote sensing at high resolution is of great use in precision pest management as it has a key role in monitoring the spatial variability.

Farmer/Producer: It is well established fact that precision pest management requires certain information and knowledge to execute the management strategies. It is therefore very important to train farmers so as to enable them to monitor and understand the dynamics of pests as well as to take appropriate decisions at a given point of time for their location.

Steps in Precision Pest Management

Basically, assessing variability and managing variability are the two important steps in precision pest management.

Assessing Variability

The assessment of in-field variability in relation to pest composition and population is a first step and plays very decisive role in making precision application of pesticides according to the assessed/ or the existing variability. The spatial variability of pests should be well recognized, adequately quantified and properly located for better and precise management. The condition maps are the crucial segments of precision pest management, which can be generated on the basis of variability of pests through (i) Surveys (ii) Point sampling and interpolation (iii) Remote sensing and (iv) Modeling.

Managing variability

Once the spatial variability of pests is established, it can further be managed via site-specific management and variable rate technology (VRT). VRT is a technique employed to apply the farm inputs at differential rates in the field in order to achieve uniform yields throughout the entire field at lower costs. Variable rate sprayers are commercially available along with other farm machineries. Variable rate sprayers are generally equipped DGPS receiver to locate the spatial variability in the field and have the ability to regulate the rate of application automatically. **In the countries like India, where economic condition of the farmers is not so good, so the concept of custom hiring can be adopted for using VRT owing to higher cost of such integrated control systems.**

Approaches

There are two main approaches namely grid sampling and management zones, which make it easy to distribute crop inputs on a spatially relative selective basis.

Grid Sampling

This approach was used to develop precision application maps for the first time. In this approach, fields are sampled at sample spacing ranging from 60-150 m along a regular grid depending on the size of field. The samples are then analyzed for desired characteristics like disease intensity, population of insect-pests and weeds as shown in **Fig. 7.1.** The results of such analysis can be interpolated to un-sampled locations by geo-statistical techniques *viz.,* Kriging and Inverse Distance Weighing (IDW) and the interpolated values are classified using GIS techniques into limited number of management zones. The boundaries of management zones are then envisaged using mapping software, which helps in developing recommendations for each interpolated zone.

Management Zones

The entire field is divided into different regions for easy management and these regions are referred to as production level management zones. This approach is considered to be more economically viable as it is easy to manage homogeneous sub-regions of a field having similar yield limiting factors. These management zones can be described using 3 GIS data layers *viz.*, bare soil imagery, topography and farmer's experience. A field can be divided into three different zones like high, medium and low based on the productivity of the area for managing the application of inputs in specific zones.

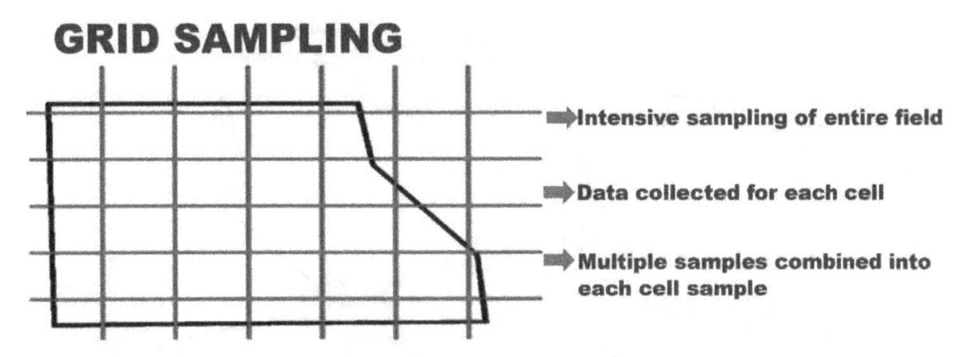

Fig. 7.1: Grid Sampling

Insect-pest Management

The decision on insecticides doses for precision management of the pests is made on the basis of pest dynamics in the management zones. Practically, it becomes sometimes/ oftenly difficult to monitor population density of insects because of their high mobility and lack of commercially available sampling techniques, which necessitates on repeated sampling thereby increasing the cost. While in precision pest management, a map of insect density will be more useful than a mean estimate of the same. Basically, precision insect management is an off-shoot of IPM, where some crop plants are maintained untreated as refugia. Similarly, such practice is also followed in precision farming. In precision pest management, proper monitoring of insect-pests is essential which could be made through precision farming technologies like RS and GIS and the management decisions on pesticidal applications can be suggested to the farmers. This type of information based technologies help in reducing the pesticide load as well as ensures better management of pests at a reduced cost.

In India, precision pest management has not yet received any attention of the researchers and the maximum amount of pesticides is applied in cotton followed by rice. Many cotton farmers commit suicide every year because of crop failure due to pest attack, which need immediate attention of the policy makers and researchers. In order to avoid such misfortune, monitoring of pest population would be the first and foremost step to decide upon the strategies. If advanced technologies like Remote sensing, GIS could not be employed, the pest population can be monitored through regular sampling. Proper spray schedule has to be developed for each management zone based on the population dynamics of the pests in respective zones. An example of melon fly monitoring through wireless sensor network is illustrated in **Figure 7.2**.

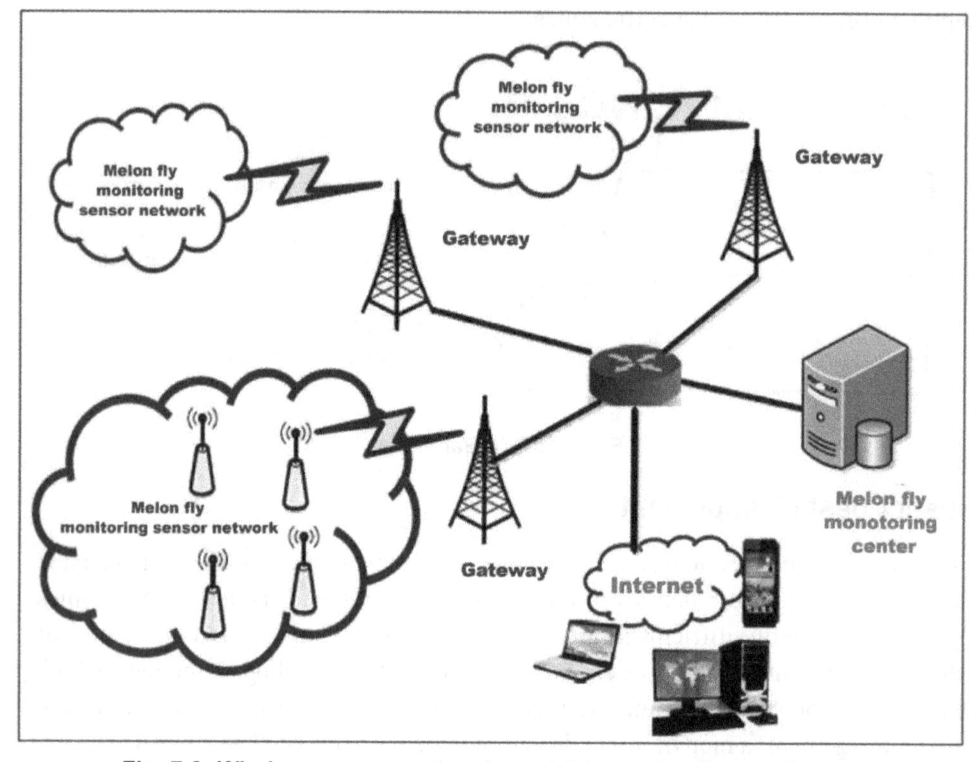

Fig. 7.2: Wireless sensor network model for melon fly monitoring

Source: Zhou and Zong, 2016

Use of Sensors for Accurate Detection of Insect-pests

Different sensors ranging from simple to the most complex ones can be used by the farmers for the detection of insect-pests in crops. The most commonly used sensors are listed below:

1. **Low-power image sensor:** It is wireless monitoring system consisting of low cost image sensors. These sensors capture images of the trapped subjects at regular intervals in the field and forward them to a control station for determination of the number of pests at each trap. So, farmers can plan their crop management strategies depending upon the information on insect population.
2. **Acoustic sensors:** These are regarded as insect-pest detection sensors, which works by detecting the noise level of the insect-pests. When the noise level

of the pests crosses the threshold limit, a sensor transmits that information to the control station for interpreting the infestation level in an area. These sensors are cost effective and serve as great tools for monitoring large field areas with very low energy consumption.

Disease Management

The disease incidence and its severity in the management zone are very important considerations to plan disease management strategy in precision farming. The basic principles and techniques employed in precision disease management are the same as followed in insect-pest management. However, monitoring of disease incidence through remote sensing demands higher resolution. Highly mobile nature of insect-pests makes it difficult to monitor their population precisely, while disease severity can easily be monitored through different surveying at regular intervals. It is also very important to understand disease epidemiology for effective and precise management of disease. Accordingly, spray schedule for a disease needs to be developed by taking into account the source of infections. Management of those diseases is very easy, which has well established forecasting models. The management strategy should include the elimination or exclusion of the pathogen since from the beginning rather than curing the disease symptoms at later or advance stages. For instance, seed treatment with specific fungicides with adequate dose is one of the best ways to manage the seed-borne diseases. It has been established that various agricultural inputs in precision farming are applied judiciously at right time in the field compared to conventional farming, hence the occurrence and severity of diseases would be much less under precision farming. Because of adoption of various agricultural processes in precision farming like proper leveling of land with lased leveller, sowing of seeds at appropriate depth through precision seeders, following advance water management techniques, applying fertilizers judiciously *etc.*, the crops in precision farming are consequently less predisposed to the disease attack. One has to give due consideration to the spatial variability of disease incidence across the crop field for precision disease management. GIS can also be used to create variability map of disease incidence for better understanding of the disease situation. Use of different methods can be effective in detection of diseases in plants for better management as depicted in **Fig. 7.3**. Similarly, importance of calendar-based strategy and the BlightPro decision support system (DSS)-based strategy for the management of late blight of potato can be differentiated very easily as shown in **Fig. 7.4**.

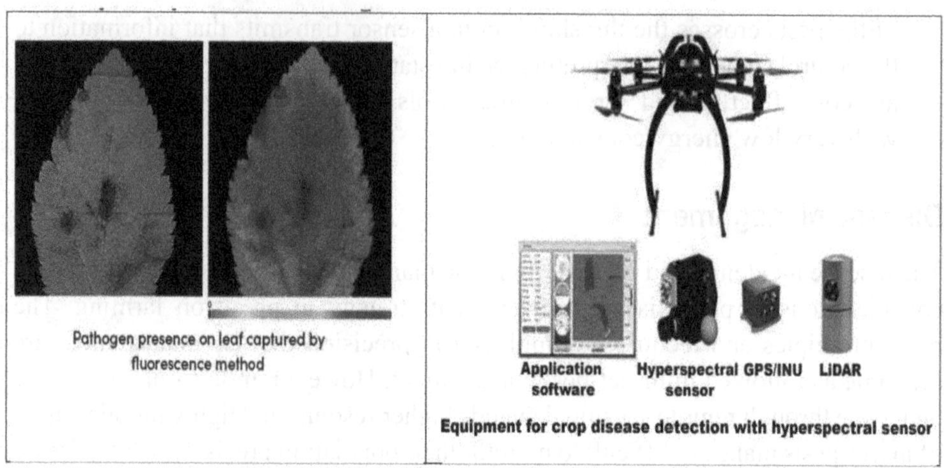

Fig. 7.3: Use of different methods for disease detection in plants

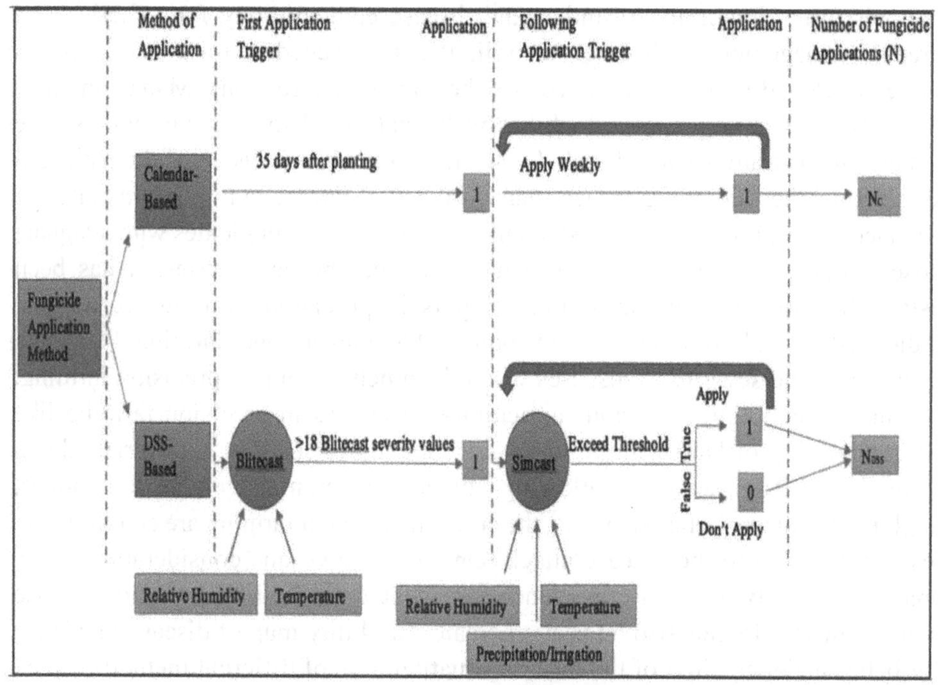

Fig. 7.4: Difference between the calendar-based strategy and the BlightPro decision support system (DSS)-based strategyThe Blitecast system reports daily severity values, which are calculated using relative humidity and temperature data as inputs (Krause *et al.*, 1975). The Simcast system reports Blight Units and Fungicide Units, which are calculated using relative humidity and temperature, as well as precipitation/irrigation data as inputs (Fry *et al.*, 1983). N stands for the number of applications for the calendar-based strategy per season per 0.41 ha and N_{DSS} stands for the number of applications for the DSS-based strategy per season per 0.41 ha.

Courtesy: Liu et al., 2017.

Sensors for the Detection of Crop Disease at Early Stages

Indirect optical sensors based on thermography, fluorescence imaging and hyperspectral techniques can be used to detect the levels of plant stress and plant instability for making interpretation on biotic and abiotic stresses as well as pathogenic diseases of the crops.

1. **Thermography disease detection technique:** It gives a general idea about disease severity under variable environmental conditions, however this technique cannot be used to identify the type of infection.
2. **Fluorescence disease detection technique:** This technique is helpful in detecting the pathogen presence by measuring changes in the levels of chlorophyll and photosynthetic activity.
3. **Hyperspectral disease detection technique:** It uses a wide range of spectrum between 350 and 2500 nm to measure plant health, which detect the changes in reflectance of some biophysical and biochemical characteristics experienced upon infection. It collects data in three dimensions, which provide more detailed and accurate information.
4. **Gas chromatography disease detection technique:** In this technique, non-optical sensors are used to determine intensity of volatile chemicals produced by the infected plants that are correlated with disease detection in crop plants. Each pathogen releases specific volatile organic compounds upon infection in plants and the sensors based on gas chromatography can accurately identify the type and nature of infection.

8

YIELD MAPPING IN HORTICULTURAL CROPS

Horticultural crops are classified as annual and perennial crops. The planting system remains stable over years, while morphological adaptation of canopy and root develops according to the environment. Temporal data over more than one season are important, since historical plant data potentially provide valuable information on the status of endogenous growth factors, *e.g.*, the status of phyto-hormones and assimilates. Horticultural products are the consequence of many manual operations and hand harvesting. In perennial fruit trees, several additional production measures like thinning of flowers and fruits, pruning are followed from time to time during the cropping period. The structures for irrigation, hail net or frost protection are the limiting factors for use of soil mapping methods in orchards, *e.g.*, for electromagnetic measurements, which are disturbed by iron installations. The specificities of the application of PA into horticultural crops in comparison to arable crops are first outlined. Then data localization and collection are executed, which is crucial and challenging in horticultural crops when various applications are addressed (**Fig. 8.1**).

Yield mapping concept: It refers to the process of collecting georeferenced data on crop yield and other characteristics like moisture content while a crop is being harvested. Various methods using a range of sensors have been developed for mapping crops yields.

Analyzing the status of canopy as well as yield mapping need low temporal resolution and can be carried out on (potentially autonomous) platforms brought to farm on certain occasions (Fig. 8.1; Table 8.1). In the other extreme, it is assumed that information on fruit is required several times during the season to follow its developmental stages. Furthermore, if detailed information on quality of produce are required, sensor signals should be collected as close to the fruit as possible to avoid unsettlement by the environment. Such data acquisition would require high manual workload. Potentially, automated stationary sensors can be implemented that provide time series of fruit data using data logger, data transfer by means of radio waves eventually using wire-less sensor network or mobile network. Performance needs for geo referencing and spatial resolution can be reduced compared to geo-referencing in remote sensing and data collection from a moving vehicle. **Table 8.1** demonstrates the examples of platforms of sensors designed to measure crop properties.

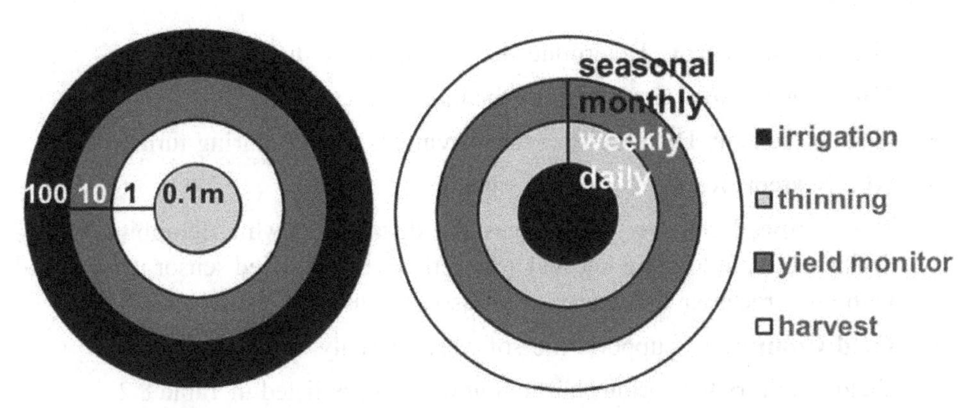

Fig. 8.1: Examples of spatial and temporal resolution of sensor data proposed for specific measures using the concept of precision horticulture

Source: Zude-Sasse *et al.*, 2016

Table 8.1: Platforms potentially carrying *in-situ* sensors commercially available for measuring crop properties in orchards on different scales

Platform	Sensor
Satellite	Infrared, Radar, multi and hyperspectral cameras
Unmanned Aerial System	Cameras (colour space, NDVI, IR, stereo), LiDAR, thermal imaging, multi or hyperspectral reading
(Autonomous) Tractor	Radar, Lidar, cameras (colour space, NDVI, IR, stereo), ultra sound, thermal imaging, multi and hyperspectral readings, yield monitor
Crane or slider on frame installation	Cameras (colour space, NDVI, IR, stereo), thermal imaging, multi and hyperspectral readings, ultra sound, LiDAR.
Stationary logger with cable or radio data transfer-eventually with wireless network	Soil sensors, climate data, balance, acoustic system, cameras, water sensors, dendrometer, optical fruit sensor.

Source: Zude-Sasse *et al.*, 2016

Yield monitor

Yield monitoring equipment was introduced in the early 90s and is considered as a conventional practice for modern agriculture. In PA as well as in precision horticulture, spatial information of the yield is pre-requisite for analysis and evaluation. Yield mapping can be carried out easily in mechanized crops with sensors added to the harvesting machine. There are some **Basic Yield Monitor Components**

- **Mass Flow Sensor**: Determine the volume to be harvested.
- **Moisture Sensor**: Compensates for moisture variability.
- **Header Sensor**: Distinguishes measurements logged during turns.
- **GPS System**: Receives satellite signals.
- **Travel Speed Sensor**: Determines the distance at which combine travels during a certain logging interval. (Sometimes travel speed sensor is measured with GPS receiver or a radar or ultrasonic sensor).
- **Field Computer**: Supports the software to analyze the process (s).

Yield monitors for various horticultural crops are listed in Table 8.2.

Table 8.2: Yield monitors for different horticultural crops.

Crop	Method of yield mapping
Citrus	Weighing pallet bins using load cells from neighbouring trees on tractor platforms. Estimating yield by tree canopy (ultrasonic sensor, LiDAR, multi-spectral camera)
Apple/ Pear / Olive	Weighing bins of handpicked fruits of neighbouring trees, geo-referenced using DGPS
Palm/Plum/Pear/ Cranberry	Numbering each tree before harvest and measuring the mass of fruits picked manually. Topographic model or local referencing
Peach/Kiwi	RFID (Radio Frequency Identification) or barcodes on the bins together with a weighing machine, RFID or barcode reader and DGPS
Potato	Load cells under the conveying chains. 2-D vision system above the conveying belt
Pecan / Broccoli	Load cells and GPS to weigh the volume and position of the platforms transferring the crop in the field on the go
Onion/Watermelon	Dividing the field into block and weighing the platforms carrying the fruits per block

Source: Zude-Sasse *et al.*, 2016

Potential Applications

- Yield maps represent the output of crop production. This information can be used to investigate the existence of spatially variable yield limiting factors. On the other hand, the yield history can be used to define spatially variable yield goals that may allow application of varying rates of inputs according to expected field productivity. The following flow chart illustrates the process; one might follow in deciding whether to invest in site-specific crop management based on yield maps. If the variability across the field cannot be explained by any spatially inconsistent field property, uniform management may be appropriate (**Fig. 8.2**). Site-specific management becomes a promising strategy if yield patterns are consistent from year to year and can be correlated to one or more field properties *e.g.* nutrient supply, topography, pest management etc.
- If the causes for yield variation are known and can be eliminated permanently, the entire area could be brought to similar growing conditions and managed uniformly thereafter.

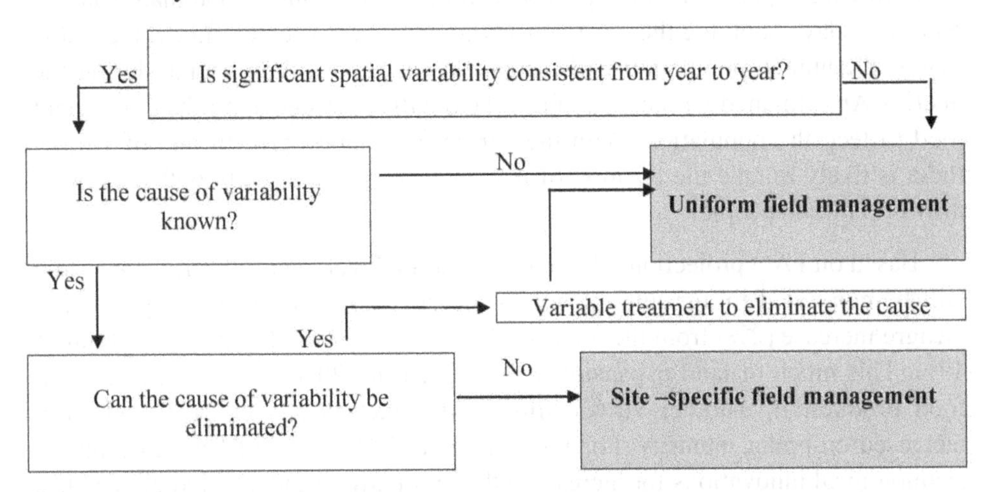

Fig. 8.2: Illustration on decision process for site-specific management

9

GREENHOUSE TECHNOLOGY: PRINCIPLES, HISTORICAL AND TECHNOLOGICAL DEVELOPMENTS

Nearly 80% of the world population will reside in urban centres by 2050 and growing population will require an estimated 100% more food than what we produce today. Applying the most conservative estimates to current demographic trends, the human population will increase by about 3 billion people during the interim. An estimated 10^9 hectare of new land will be needed to produce sufficient food to feed the population. With the current population growth rate of 1.02%, India is likely to take the position of *numero uno* by the end of 2030 with more than 1.53 billion people.

Based on FAO projections, 13% more land in developing countries like India will be converted to agricultural use in the next 30 years, which appears to a meagre increase (2%) from the 38% of global land area used in 2008 to a total of 40%. This much of land expansion will only ensure 20% of future increases in food production. Another 10% additional production can be projected from increased cropping intensity. For the remaining 70%, we will have to think of technological innovations for increasing the productivity and also judiciously use the ones at hand. In addition to the fact that land is limited and reclamation is a slow process often coupled with environmental degradation, we are also losing land at an alarming rate due to climate change and desertification.

According to the USDA Economic Research Service, the development of new agricultural technologies like advances in genetics, nutrition, disease and pest control and livestock management was an important factor in the 20th century for making improvements in the productivity of various crops. As land resource is finite and allocation of this resource further complicates the problem. There is also a need to minimize the environmental effects in relation to emission of greenhouse gases, degradation of soil and water resources, and biodiversity. The climate change on a global scale will definitely affect agriculture and consequently the world's food supply. Climate change *per se* is not necessarily harmful, but the problems arising from extreme events are difficult to predict, which may have direct impact on agricultural production. Erratic behaviour of rainfall and uncertainty in temperature will consequently reduce crop productivity. Developing countries in the tropics will be more vulnerable to such changes enforced by climate change. Latitudinal and altitudinal shifts in ecological and agro-economic zones, land degradation, extreme geophysical events, reduced water availability, and rise in sea level and salinization have been postulated in a FAO report.

In view of such anticipations, we need to deploy such agricultural technologies that have nearly neutral or positive impact on the environment. The problems of protecting the environment and balancing the world's need for energy and food require comprehensively multifarious approach. The scope for horizontal expansion is very little and the only available alternative is vertical expansion by increasing productivity and cropping intensity with the use of modern methods of faming such as protected farming/vertical farming employing plant environment control measures, use of good quality seeds, inputs management (fertilizer, irrigation *etc.*) and plant protection. Protected farming is economically more rewarding for the production of high value-low volume crops, seeds and planting materials *etc*. The constraints of environment prevalent in a particular region can be overcome by cultivating crops under appropriate structures designed for achieving year round cultivation with increased productivity by 25-100% and in certain cases even more, as well as conservation of irrigation water by 25-50%. Protected farming is regarded as an alternate farming method with much higher carrying capacity. Vertical farming holds the promise of addressing these issues by enabling more food to be produced with fewer resources.

Primarily, greenhouse is a concept of manipulating the crop micro-environment and it is generally a house like structure where the solar radiations, temperature, humidity, CO_2 can be manipulated appropriately to meet out the requirements of a particular crop. Greenhouse technology is the most intensive form of commercial cultivation and could well be called as 'food factories". With the advent of agricultural technologies, more emphasis is now being laid on quality production of various agricultural commodities along with quantity to meet out ever increasing food requirements of masses. But due to erratic behaviour of weather, the crops grown in open conditions are generally exposed to varying levels of temperature, humidity, wind flow *etc*., which ultimately influence the productivity of a crop extensively. Greenhouse technology has emerged as one of the potential technologies to overcome climatic diversity efficiently and simultaneously fulfills the requirements of growers in terms of manifold increase in yield as well as quality of the produce as per the demand of the market. In recent times, protected cultivation has allowed cultivation of high valued crops in off-season with high yield and better quality. This technology offers an opportunity to manage different climatic parameters as per the requirement of a crop and protect crops from different weather vagaries. Greenhouse technology is one of the most important applications of precision farming for achieving the objectives of protected agriculture/horticulture, wherein growing environment inside the protected structure can partially or fully be controlled by using structural engineering principles to realize optimum growth and yields in different crops.

What is Greenhouse?

A greenhouse refers to a framed structure covered with transparent and translucent material and supported by galvanized iron, bamboo or/ any other wooden material, wherein an enclosed area is used for crop cultivation under partially or fully controlled conditions. Greenhouse is a broad term commonly used to endorse greenhouse technology, derived from the universal working principle *i.e.* Greenhouse effect. The cladding/covering materials in greenhouse selectively allow solar radiations of short wavelengths to pass through it, which falling on ground are radiated back as long wavelength radiations and trapped as thermal energy by cladding material inside the greenhouse. This phenomenon is known as "Greenhouse Effect" (**Fig. 9.1**). Specific terms are used to refer a greenhouse as glasshouse, polyhouse, polycarbonate house, shade net house when cladding material used in a protected structure is glass, polyethylene or plastic films/sheet, carbonate sheet, agro shade net, respectively. Specifically, where the covering material is glass, the structure may be referred to as a 'glasshouse'. A 'greenhouse' or 'polyhouse' refers to the use of plastic films or sheeting.

Selection of such protected structures should be done as per the suitability and designs recommended to specific locations to harness best of greenhouse technology.

Greenhouse Effect

- The greenhouse effect is a natural warming process of the earth.
- When the solar energy reaches the earth's surface, some of it is reflected back to space and rest is absorbed.
- The absorbed energy warms the earth's surface, which then emits heat energy back towards space as long wave radiations.
- The outgoing long wave radiations are partially trapped by greenhouse gases such as carbon dioxide (CO_2), Methane (CH_4), Nitrous oxide (NO_2), chlorofluorocarbons (CFCs), Sulfur hexafluoride and water vapours, which

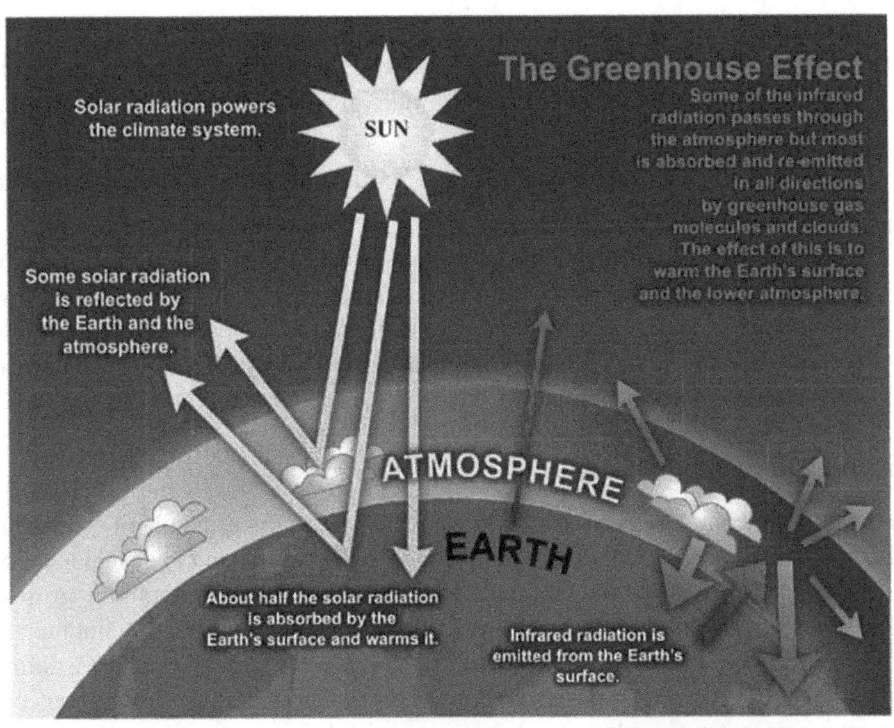

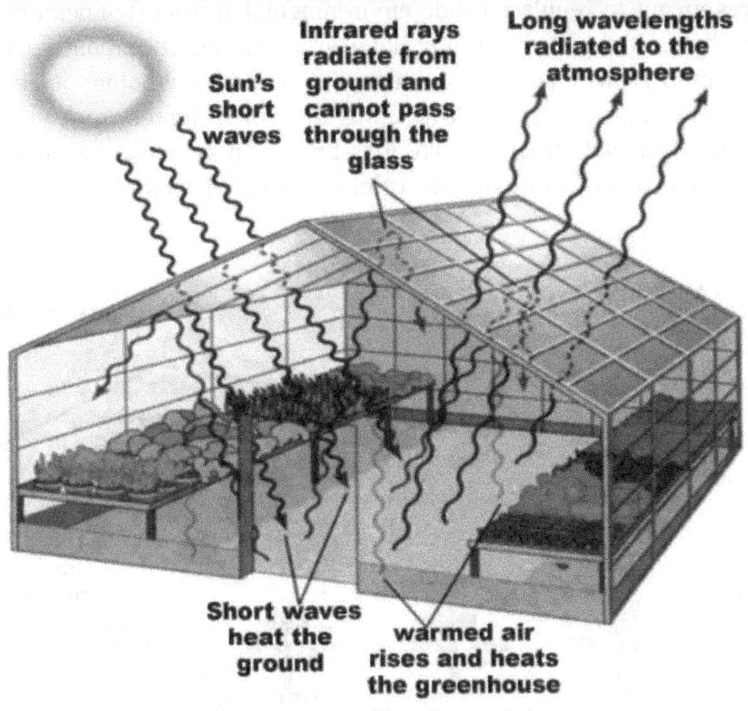

Fig. 9.1: Greenhouse effect

then radiate energy in all directions, thus warming the earth's surface and atmosphere. CO_2 accounts for 50% of the greenhouse effect, methane for 20%, CFCs for 14% and the remaining proportion is governed by other components including water vapours.

- Without these greenhouse gases, the earth's average surface temperature would be about 33°C cooler.
- This phenomenon is used in greenhouse technology to optimize growing conditions suitably matching to the requirement of crop plants for better growth and development under protected environment *i.e.* protected cultivation. This technology has now made it possible to grow crop plants out of the season (off-season) and throughout the year.

Greenhouse Technology

Greenhouse technology is a technology which employs the principle of greenhouse effect with the involvement of structural engineering component to design location-specific protected structures for creating optimum/ near to optimum growing conditions and technological approaches for proper growth and development of crop plants inside such structures. Therefore, greenhouse technology can be defined as the technology which uses structural engineering aspects to design protected structures meant to regulate inside environmental factors (temperature, relative humidity, light, CO_2) partially or fully along with some specific production techniques for better crop growth and development. Greenhouse technology is also referred to protected cultivation as it allows cultivation of different crops under certain specifically designed structures like glasshouses, polyhouses, shade net houses *etc.* protecting them from extremes of environmental factors (**Fig. 9.2**).

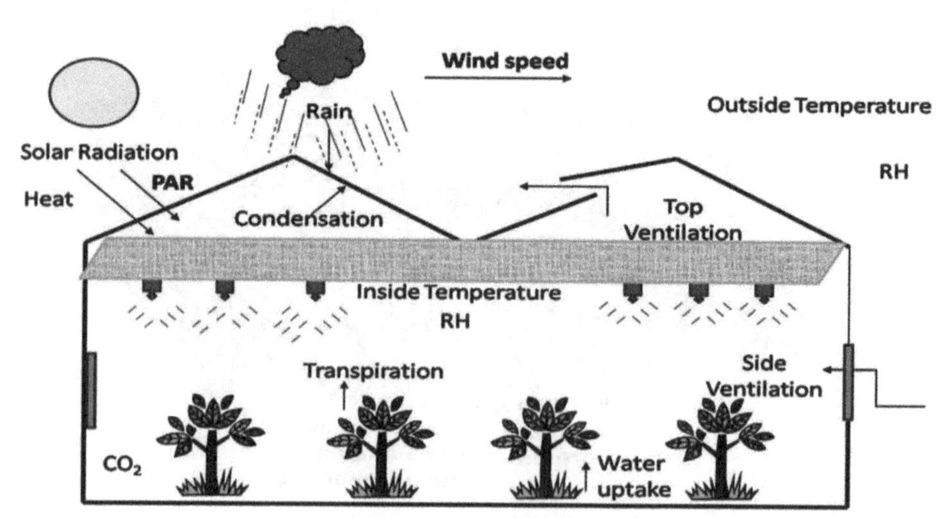

Fig. 9.2: Greenhouse Technology

Global Outlook of Protected Cultivation

- First documentation on the use of protected structures with transparent slate-like plates or sheets of mica or alabaster as cladding material for crop production during the sovereignty of Emperor Tiberius Caesar, a Roman Empire having movable beds of cucumbers for keeping outside during favourable and inside during inclement weather (14-37 AD).
- Protected structures of crude square or rectangular wooden or bamboo frames or structures covered with panes of glass, oiled paper, or glass bells to cover hotbeds appeared mainly in England, Holland, France, Japan, and China for off-season cultivation of vegetables and small fruits (15th to 18th centuries).
- Greenhouses with glass on one side only in the form of a sloping roof came into existence. Stoves and heating flues came in use for first time in glasshouses (late 1600s to 1700s).
- Use of glass for the front and on both ends of lean-to type protected structures. The two sided greenhouses emerged in the 1800s, and England, Holland, France and the Scandinavian countries were the leaders of early developments. Off-season cultivation became popular particularly in England, and in the United States, New York and Boston started off-season production of vegetables initially in hot beds and then under greenhouses during 1700s to 1800s.
- Greenhouse products like grapes, melons, peaches and strawberry started to appear in English markets (By the middle of the 1800s).
- v In America, lettuce and radish were the first greenhouse crops followed by cucumbers and tomatoes. Protected cultivation expanded from Boston area Grand Rapids, Mich., to Cleveland and Toledo, Ohio. By the end of century, approximately 1000 greenhouses were functional in the United States for off-season cultivation of winter vegetables (1860s).
- Tomatoes became very popular greenhouse crop (1870s to 1880s).
- Commercial activities were established for greenhouse crop production system. The technology of greenhouses and forcing frames later extended from Europe to America and other countries. In urban and peri-urban area of China, Japan and Korea, glasshouses with glass only on the roof and southern wall were built, while the northern and side walls were constructed of either concrete or adobe embanked with bales of rice straw for insulation. These structures were in use for cultivation of cucumber, tomato, eggplant and beans. Many of the structures were not heated, but covered with straw mats at night (By the end of the 19th century).
- Vegetable cultivation under glasshouses of 550 ha in the United States with 43% representation by tomatoes, 33% by cucumbers and remaining by others.

Flowers and ornamentals cultivation occupied nearly double the area of vegetable crops (By 1929).

- After the 2nd world war, greenhouse technology got acceleration in development especially in Western European countries, Netherlands being the leader. Agro-technical systems, aeration solutions and accompanying accessories were added to the protected structures, while the foundations were improved to heavy steel constructions covered by rigid glasses.
- Meanwhile, other areas particularly Asian and Mediterranean countries witnessed a phenomenal increase in greenhouse culture of high valued vegetable crops with the use of plastic for non-heated greenhouse. Mediterranean region observed expansion through plasticulture allocations, but experienced a gradual transition from the production of food crops to flowers, potted, bedding plants and ornamentals (1950s).
- Glasshouse construction in the United States and Canada was at its peak. Concurrently, greenhouse development was taking place for vegetables majorly tomatoes in North-Western Europe. Cut flower and ornamental plant industry was also emerging particularly in England, Germany, Holland, and Scandinavia. By 1960, Holland occupied significantly maximum area (5000-6000 ha) under glasshouses with highest cultivation of tomato (75%). Denmark had 600 ha under glasshouses with 10% growth every year mostly devoted to cucumber and tomato production. Belgium also started cultivating table grapes in small units in the 1950s and the early 1960s. Simultaneously, greenhouse industry started emerging in Michigan with maximum focus on rhubarb cultivation in heated houses during winter and celery cultivation under row cover with parchment paper (late 1950s and early 1960s).
- ***Comite International des Plastiques en Agriculture*** **(CIPA):** International Committee for Plastics in Agriculture was founded in 1959. First Congress was held on 5th May 1964, Avignon (France) and the committee was registered in accordance with French law on 18th July, 1969. Later on, it's headquarter was relocated to Spain on 1st Jan 2007. Previously also known under the English acronym *ICPA.* Objective of organization is to liaise between the various established National Committees (NCPA); promote the formation of new committees; provide information to producers, converters and users worldwide on research and studies relating to the use of plastic materials in agriculture; encourage in the widest economic sense, provide solutions to scientific and technical problems in the field of plastics in agriculture through international harmonization; promote international standardization of plastics in agriculture and their experimentation in order to encourage the creation of quality standard, subjected to CIPA assessment; study and analyse problems relating to plastics in agriculture and the environment.

- A spectacular shift was observed in greenhouse industry with the emergence of hydroponic culture especially for tomatoes- an attractive and expensive venture in the United States. There was more concentration on greenhouse cultivation of flowers, potted plants, ornamental and bedding plants. Drip irrigation development took place in Israel which travelled to the United States in 1970 (Late 1960s and early 1970s).
- The area under greenhouses (glass, fibre glass and plastic) for high value food crops was estimated to be 150000 ha. In addition, about 50000 ha area was exclusively meant for the production of flowers, ornamental and bedding plants (By 1980).
- There was remarkable increase in area under protected cultivation in the Westlands and elsewhere in Holland for round the year production of high quality vegetable crops like tomato, pepper, eggplant, cucumber, lettuce, cut flowers (rose, chrysanthemum, carnation, tulip, lily) and potted plants (Ficus, dracaena, kalanchoe, begonia, azalea). Carbon dioxide enrichment along with other high technology inputs (differences in day and night temperature), artificial substrates (media), bottom heat and precise control of inputs like water, temperature, and nutrients were in use to manage desirable micro-climate around the plants for higher output (1990s).

Indian Outlook of Protected Cultivation

- In India baring the pioneering research and development work on protected cultivation of vegetables by **Defence Research and Development Organization (DRDO) during 1960s at its one of the establishments namely, Field Research Laboratory (FRL), Leh now known as Defence Institute of High Altitude Research (DIHAR), use of greenhouse technology started only during 1980s.** Initially, it was mainly practiced as research activity. This may be because of India's emphasis so far had been on achieving self-sufficiency in food grain production.
- Indo-American Hybrid Seeds (India), Bangalore, is the pioneer in India to make use of greenhouse technology since 1965 for commercial productions of flower seeds.
- In view of the globalization of international market and tremendous boost and fillip that is being given for export of agricultural produce, there has been a spurt in the demand for greenhouse technology. **The National committee on the use of Plastics in Agriculture (NCPA) was constituted in March, 1981** by Government of India under the Ministry of Petroleum Chemicals and Fertilizers. NCPA started to work under the Department of Agriculture and Cooperation with effect from 1993. Initially, the committee was set up

for two years term but was later extended for another two years up to 6th March, 1995. Further, the committee was reconstituted in 1996 in order to make it more effective and to focus its endeavor in a coordinated manner for promoting the applications in Horticulture. **The committee was reconstituted as National Committee on Plasticulture Applications in Horticulture (NCPAH) in 2001 with Head Quarter at New Delhi.**

Mandate of NCPAH

- Coordination and promotion of horticulture/agriculture development through the use of plastics in agriculture (plasticulture) with special reference to harnessing available natural resources such as water and sunlight in improving the productivity and quality of the produce.
- Recommendation of suitable policy measures for the promotion of plasticulture in the country.
- Facilitating increased adoption of various plasticulture applications like drip and sprinkler irrigation systems, protected cultivation, community tanks, post harvest management *etc.*
- Facilitating and strengthening research and development and to build data base in plasticulture applications.
- Development of appropriate quality standards for the products used in plasticulture and ensure proper adoption of such standards in the field.
- Suggesting suitable ways and means for effective implementation of centrally sponsored schemes on micro-irrigation having integration with National Horticulture Mission (NHM), Mission for Integrated Development of Horticulture (MIDH), Technology Mission for Horticulture in the North East (TMNE), *Rashtriya Krishi Vikas Yojna* (RKVY) *etc.* in relation to plasticulture components in these schemes.
- Supervising and monitoring the performance of **Precision Farming Development Centres (PFDCs)** for overall development of precision farming technologies and hi-tech interventions in the country.

Precision Farming Development Centres

Precision Farming Development Centres (PFDCs) were established in India to promote "Precision Farming and Plasticulture Applications in high-tech horticulture" and are located in State Agricultural Universities (SAUs), ICAR Institutes such as IARI, New Delhi; CIAE, Bhopal; CISH, Lucknow and IIT, Kharagpur. These centers have been operating as hub-centers of plasticulture and precision farming in respective states. NCPAH during the year 2008-09 established five new

Precision Farming Development Centres at Bhopal, Imphal, Leh, Ludhiana and Ranchi under the centrally sponsored scheme on micro-irrigation. The details of different PFDC are depicted in the map below.

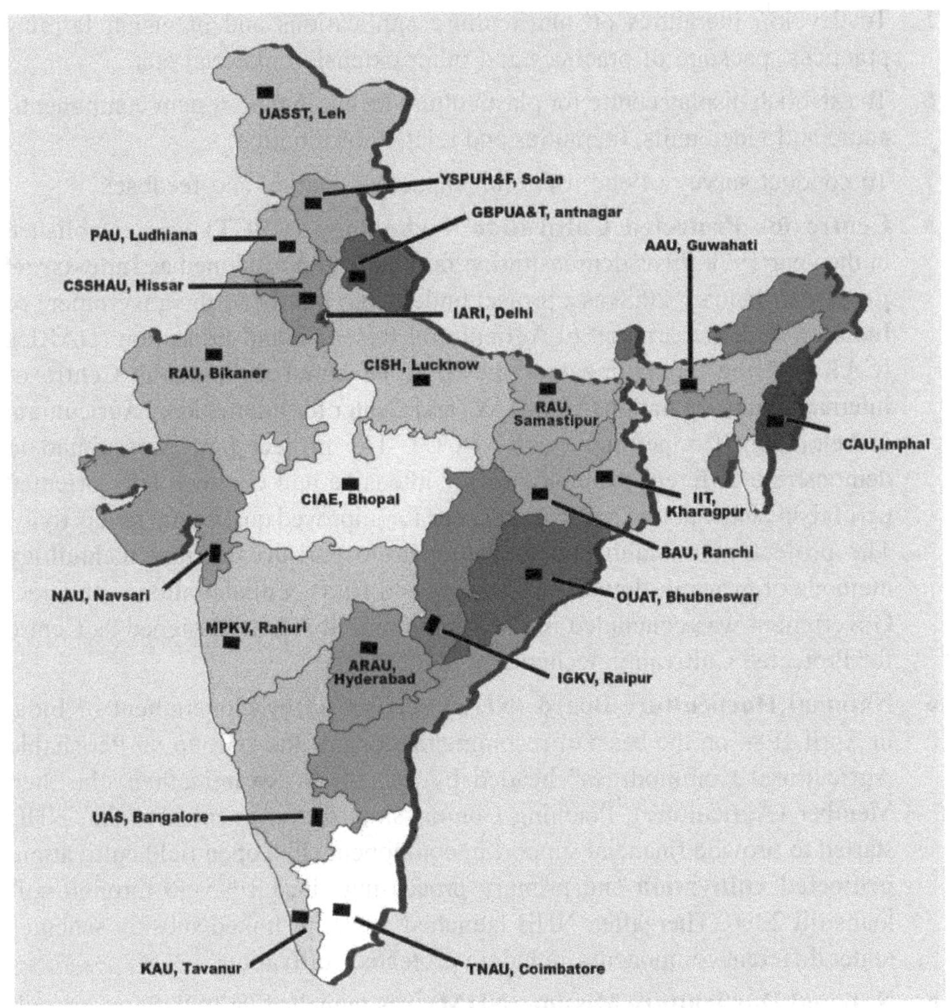

Precision Farming Development Centres, India

Courtesy: www.ncpahindia.com

Mandate of PFDC

1. To undertake trials and experiments on plasticulture applications for the development of crop specific plasticulture technologies.
2. To provide technical input to the Sate Govt. in implementation of micro-irrigation, NHM and other related schemes.

3. To demonstrate proven plasticulture technologies at PFDC and farmers' field.
4. To transfer technologies through training and awareness programmes, participation in agricultural related events, print and satellite media.
5. To develop literatures on plasticulture applications and precision farming practices, package of practices and other extension material etc.
6. To establish display centre for plasticulture technologies, system components, audio and video units, literatures and related information.
7. To conduct survey of end users for impact evaluation and feedback.

- **Centre for Protected Cultivation Technology (CPCT)** was established in the year 1998-99 as demonstration farm and commissioned as **Indo-Israel project** in January 2000 as a project undertaken jointly by the government of India through Department of Agricultural Research and Education (DARE), ICAR and the Government of the State of Israel through the Centre of International Cooperation (MASHAV) and Center for International Agricultural Development Cooperation (CINADCO). The project farm was aimed to demonstrate different technologies for intensive and commercially oriented peri-urban cultivation of horticulture crops for improved quality and productivity. The project was established to demonstrate peri-urban, high technology methods of growing flowers, vegetables and fruits. Collaboration with Israel Government was concluded in 2004 and the unit was re-designed as Centre for Protected Cultivation technology (CPCT).
- **National Horticulture Board (NHB)** was set up by Government of India in April 1984 on the basis of recommendations of the "Group on Perishable Agricultural Commodities" headed by Dr. M. S. Swaminathan, the then Member (Agriculture), Planning Commission, Government of India. NHB started to provide financial support in components like open field cultivation, **protected cultivation** and primary processing since 1995-96 through soft loans till 2000. Thereafter, NHB launched its credit linked subsidy schemes under different components including protected cultivation.
- **National Horticulture Mission (NHM)** was launched in 2005-06 as a result of which significant progress has been made in horticulture crops through various interventions including protected cultivation.
- ***Rashtriya Krishi Vikas Yojana*** was initiated in 2007 as an umbrella scheme for ensuring holistic development of agriculture and allied sectors by allowing states to choose their own agriculture and allied sector development activities as per the district/state agriculture plan. Based on feedback received from States, experiences garnered during implementation in the 12th FYP Plan and the inputs provided by stakeholders, RKVY guidelines have been revamped as **RKVY – RAFTAAR – Remunerative Approaches for Agriculture**

and Allied sector Rejuvenation to enhance efficiency, efficacy and inclusiveness of the programme. Creation of value resources and protected cultivation (Greenhouse/ Polyhouse/Shade net house infrastructures) is one of the components of infrastructure development under Horticulture Sector.

- **Mission for Integrated Development of Horticulture (MIDH)** is a Centrally Sponsored Scheme being implemented *w.e.f.* 2014-15 for holistic growth of the horticulture sector covering fruits, vegetables, root and tuber crops, mushrooms, spices, flowers, aromatic plants, coconut, cashew, cocoa. MIDH subsumed ongoing missions/schemes of the Ministry namely National Horticulture Mission (NHM), Horticulture Mission for North East & Himalayan States (HMNEH), National Horticulture Board (NHB), Coconut Development Board (CDB) and Central Institute for Horticulture (CIH), Nagaland. All States including North Eastern States and UTs are covered under MIDH.

  Major Interventions of MIDH

  a) Setting up of nurseries, tissue culture units for the production of quality seed and planting material

  b) Area expansion *i.e.* Establishment of new orchards and gardens for fruits, vegetables, and flowers

  c) Rejuvenation of unproductive, old, and senile orchards

  d) Protected cultivation *i.e.* polyhouse, greenhouse, net house *etc.* to improve the productivity and grow off-season high value vegetables and flowers

  e) Organic farming and certification

  f) Creation of water resources structures and watershed management

  g) Bee keeping for pollination management

  h) Horticulture mechanization

  i) Creation of Post Harvest Management and Marketing infrastructure.

- **Horticulture Mission for North East and Himalayan States (HMNEH)** was launched by the Government of India in order to improve livelihood opportunities and bring prosperity to the North Eastern Region (NER) including Sikkim in 2001-02. Considering the potential of Horticulture for socio-economic development of Jammu & Kashmir, Himachal Pradesh and Uttarakhand, Technology Mission was extended to these States during 2003-04. HMNEH is based on the "end-to-end approach" taking into account the entire gamut of horticulture development with all backward and forward linkages in a holistic manner. Government of India (GOI) has approved the implementation of Technology Mission for Integrated Development of Horticulture in North Eastern States including Sikkim, Jammu & Kashmir, Himachal Pradesh and

Uttarakhand (**TMNE**) during XI FYP Plan. As per the approval of GOI, the scheme will be implemented in the name of HMNEH.

Pradhan Mantri Kaushal Vikas Yojana (PMKVY)

The Union Cabinet approved India's largest Skill Certification Scheme, *Pradhan Mantri Kaushal Vikas Yojana* (PMKVY) on 20th March, 2015. The Scheme was subsequently launched on 15 July, 2015 on the occasion of World Youth Skills Day by Honorable Prime Minister, Shri Narendra Modi. PMKVY is implemented by National Skills Development Corporation (NSDC) under the guidance of the Ministry of Skill Development and Entrepreneurship (MSDE). With a vision of a "Skilled India", MSDE aims to skill India on a large scale with speed and high standards and the PMKVY is a flagship scheme meant to realize this vision. Owing to the its successful first year of implementation, the Union Cabinet has approved the Scheme for another four years (2016-2020) to impart skill to 10 million youths of the country. Agriculture Skill Council of India (ASCI) is working under the aegis of Ministry of Skill Development & Entrepreneurship (MSDE) towards capacity building by bridging gaps and upgrading skills of farmers, wage workers, self-employed and extension workers engaged in organized/ unorganized activities of Agriculture & Allied Sectors though different qualification packs including **protected cultivation**. ASCI in its endeavor to develop and upgrade the skills of Indian agriculture & agri-allied workforce, is collaborating with all key stakeholders including Agricultural and Veterinary Universities, and Research Institutions.

Agriculture / Horticulture Universities

1. Punjab Agriculture University (PAU), Ludhiana
2. Tamil Nadu Agriculture University (TNAU), Coimbatore
3. Orissa University of Agriculture & Technology (OUAT), Bhubaneswar
4. University of Horticultural Sciences (UHS), Bagalkot
5. Indira Gandhi Krishi Vishwavidyalaya (IGKV), Raipur, Chhattisgarh
6. University of Agricultural & Horticultural Sciences (UAHS), Shimoga, Karnataka
7. Central Agricultural University (CAU), Manipur
8. Professor Jayashankar Telangana State Agricultural University (PJTSAU), Hyderabad
9. Chaudhary Charan Singh Haryana Agricultural University (CCSHAU), Hisar
10. Banda University of Agriculture & Technology (BUAT), Banda

11. Maharana Pratap University of Agriculture & Technology (MPUAT), Udaipur
12. Navsari Agricultural University (NAU), Navsari, Gujarat
13. Bihar Agriculture University (BAU), Sabour, Bhagalpur
14. G B Pant University of Agriculture & Technology (GBPU&AT), Pantnagar
15. Birsa Agricultural University (BAU), Ranchi
16. Dr. YSR Horticultural University (Dr. YSRHU), Andhra Pradesh
17. Dr. Yashwant Singh Parmar University of Horticulture & Forestry (YSPUHF), Solan, Himachal Pradesh
18. University of Agricultural Sciences (UAS), Dharwad
19. Kerala Agricultural University (KAU), Kerala
20. Jawaharlal Nehru Krishi Vishwa Vidyalaya (JNKVV), Jabalpur
21. Rajmata Vijayaraje Scindia Krishi Vishwa Vidyalaya (RVSKVV), Gwalior
22. CSK Himachal Pradesh Krishi Vishwavidyala (CSK HPKV), Palampur
23. Bidhan Chandra Krishi Vishwavidyalaya (BCKV), West Bengal
24. Acharya N.G. Ranga Agricultural University (ANGRAU), Guntur
25. Swami Keshwanand Rajasthan Agricultural University (SKRAU), Bikaner
26. Sardar Vallabhbhai Patel University of Agriculture & Technology (SVPUAT), Meerut
27. Junagadh Agricultural University (JAU), Junagardh
28. Sri Karan Narendra Agriculture University (SKNAU), Jobner
29. Assam Agricultural University (AAU), Jorhat
30. Maharana Pratap Horticultural University (MHU), Karnal
31. Narendra Deva University of Agriculture & Technology (NDUAT), Faizabad
32. University of Agricultural Sciences, Raichur

Veterinary/Animal Sciences/ Fisheries Universities

1. Guru Angad Dev Veterinary & Animal Sciences University (GADVASU), Ludhiana
2. Tamil Nadu Veterinary and Animal Sciences University (TANUVAS), Chennai
3. Tamil Nadu Fisheries University (TNFU), Nagapattinam
4. West Bengal University of Animal & Fishery Sciences (WBUAFS), Kolkata
5. U. P Pandit Deen Dayal Upadhyaya Pashu Chikitsa Vigyan Vishwavidyalaya Evam Go Anusandhan Sansthan (DUVASU), Mathura, Uttar Pradesh
6. Maharashtra Animal and Fishery Sciences University (MAFSU), Nagpur

7. Kerala Veterinary & Animal Sciences University (KVASU), Kerala
8. Sri Venkateswara Veterinary University (SVVU), Tirupati
9. Lala Lajpat Rai University of Veterinary & Animal Sciences, Hisar
10. Rajasthan University of Veterinary & Animal Sciences (RAJUVAS), Bikaner
11. Nanaji Deshmukh Veterinary Science University (NDVSU), Jabalpur
12. The National Dairy Research Institute (NDRI), Karnal
13. Karnataka Veterinary, Animal & Fisheries Sciences University (KVAFSU), Karnataka
14. Central Institute of Fisheries Education (CIFE), Mumbai
15. Chhattisgarh Kamdhenu Vishwavidyalaya (CGKV), Anjora
16. Bihar Animal Sciences University (BASU), Patna
17. Kerala University of Fisheries & Ocean Studies (KUFOS), Kochi

CSIR Institutes

1. CSIR- National Institute of Oceanography (CSIR-NIO), Goa
2. CSIR- Institute of Himalayan Bioresource Technology (CSIR-IHBT), Palampur
3. CSIR- Indian Institute of Toxicology Research (CSIR-IITR), Lucknow (Nov)
4. CSIR- Central Scientific Instruments Organization (CSIR-CSIO), Chandigarh
5. CSIR- National Botanical Research Institute (CSIR-NBRI), Lucknow
6. CSIR- Indian Institute of Integrative Medicine (CSIR-IIIM), Jammu
7. CSIR- National Institute for Interdisciplinary Science & Technology (CSIR-NIIST), Thiruvananthapuram
8. CSIR- Advanced Material & Processes Research Institute (CSIR-AMPRI), Bhopal

Skill Universities

1. Haryana *Vishwakarma* Skill University (HVSU), Haryana

Open University

1. Indira Gandhi National Open University (IGNOU), New Delhi

AICRP on Plasticulture Engineering and Technology

The All India Coordinated Research Project on Plasticulture Engineering and Technologies (PET) become operational in 1988 during VII FYP Plan period (known as AICRP on Application of Plastics in Agriculture) to undertake research and extension activity pertaining to water management, **protected farming**, post-harvest produce management *etc*. During XII FYP Plan, the project became operative at fourteen Centres located in different agro-ecological regions with its coordinating unit located at ICAR-CIPHET (Central Institute of Post Harvest Engineering and Technology), Ludhiana.

Mandate

- To develop strategies for the use of plastics in agriculture with major emphasis on surface covered cultivation, lining of ponds for rainwater harvesting, storage of water, micro-irrigation systems, packaging and storage, transportation of agricultural produce and products, aquaculture and livestock management.

Objectives

- To apply plastics in agriculture both in production agriculture and post harvest management
- To identify newer areas for plastic applications in agriculture particularly in inland fisheries and animal shelters as well as environment control
- To carry out operational research on proven technologies at pilot level with area saturation approach
- To disseminate plasticulture technologies through publications, media, exposure and training programmes, workshops, developing linkages with industry, other stakeholders and catalyzing developmental programmes

Thrust areas

- Standardization of location specific designs of polyhouses and shade net houses for round the year use.
- Development/improvement of laying techniques and its mechanization for plastic mulch and low tunnels.
- Techniques for rainwater harvesting in plastic lined ponds and their management.
- Development of plastic based low-cost pressurized irrigation system equipped with pumping unit powered by non-conventional energy source.
- Development of plastic devices/systems for intensive fish culture and animal shelter.
- Application of plastics for post harvest management (handling, storage and packaging) of important agricultural produce.
- Development/improvement of plastic components of farm equipment.

Presently this AICRP has fourteen Centres as listed below:

List of Cooperating Centres

1. Vivekananda Parvatiya Krishi Anusandhan Sansthan (VPKAS), Division of Crop Production, Almora 263 601 (Uttarakhand) operating since 1988.
2. Central Institute of Freshwater Aquaculture (CIFA), PO: Kausalayaganga, Bhubaneswar 751 002 (Orissa), operating since 1988.
3. Punjab Agricultural University (PAU), Dept. of Soil & Water Conservation Engg., College of Agril. Engg. Ludhiana 141 004 (Punjab) operating since 1988.
4. Central Institute of Post Harvest Engineering and Technology (CIPHET), Malout Hanumangarh Byepass Road, Abohar 152 116 (Punjab) operating since 1990.
5. Junagarh Agricultural University (JAU), Department of Renewable Energy, College of Agril. Engg & Tech, Junagadh 362 001 (Gujarat) operating since 2004.
6. Birsa Agricultural University (BAU), Department of Agricultural Engineering, PO Kanke, Ranchi 834 006 (Jharkhand) operating since 2004.
7. Sher-e-Kashmir University of Agricultural Sciences and Technology of Kashmir (SKUAST-K), Dept. of Agril. Engg., Shalimar Bagh, Srinagar 191 121 (J & K) operating since 2004.
8. Maharana Pratap University of Agriculture and Technology (MPUAT), College of Technology and Engineering, Department of Soil Water Engg., Udaipur 313 001 (Rajasthan) operating since 2009.
9. Dr. Balasaheb Sawant Konkan Krishi Vidyapeeth (DBSKKV), Department of Farm Structure, College of Agricultural Engineering and Technology, Dapoli, District Ratangiri 415712 (Maharashtra) operating since 1st April, 2015.
10. Central Institute for Research on Goats (CIRG), Makhdoom, P.O. Farah 281122, Mathura (Uttar Pradesh) operating since 1st April, 2015.
11. University of Agricultural Sciences (UAS), College of Agricultural Engineering, Raichur 584102, (Karnataka) operating since 1st April, 2015.

North Eastern Region

12. ICAR Research Complex for NEH Region, Div. of Agril. Engg., Umroi Road, Umiam 793 103 (Meghalaya) operating since 2000.
13. Central Agricultural University (Imphal), College of Agriculture Engineering and Post Harvest Technology (CAEPHT), Ranipool 737 135, Gangtok (Sikkim) operating since 2009.

14. National Research Centre on Yak, Dirang 7900101, District West Kameng (Arunachal Pradesh) operating since 1st April, 2015.

Advantages of Protected Cultivation

- Provides favourable micro-climatic conditions for growth and development of plants
- Cultivation in all seasons is possible
- Higher yield with better quality per unit area
- Conserves moisture thus needs less irrigation
- More suitable for cultivating high valued / off - season crops
- Helps to control pests and diseases
- Helps in hardening of tissue cultured plants
- Helps in raising early/off-season nurseries
- Round the year propagation of planting material is possible
- Protects the crops from wind, rain, snow, bird, hail *etc.*
- Generates self- employment opportunities for educated youth

Technological challenges

1. Non-availability of indigenously bred/ developed varieties/ hybrids for protected cultivation
2. High initial and operational cost of protected structures.
3. Lack of work on development/ standardization of designs of protected structures for different agro-climatic regions of the country
4. Very little work on the development of POPs, INM and IPDM modules for greenhouse crops.
5. Non-availability of tools and implements facilitating crop production operations under greenhouse. For instance, electric vibrator for pollination in greenhouse tomatoes.
6. Very little or negligible work on development of techno-economic feasible and region-specific cropping schemes.
7. There is a problem of pollination in greenhouse crops
8. Increasing threat of bio-stresses in greenhouse cultivation particularly root-knot nematodes and *Fusarium* wilt.

Strategies to strengthen protected cultivation in India:

1. Establishment of National Research Centre on Protected Cultivation
2. Development of suitable varieties/ hybrids for protected cultivation
3. Identification of new and potential crop types for protected cultivation
4. Development of cost-effective and location-specific designs of greenhouses
5. Development and use of new generation biodegradable polymers
6. Designing and simulation of photovoltaic greenhouse (PVG) systems for Indian conditions
7. Adoption of novel technologies in cladding material like NIR reflecting, Cool plastic film, photo-selective film that creates cooler temperature in the greenhouses
8. Use of antivirus film *i.e.* UV block film that completely blocks UV radiation to adversely affect thrips and whitefly
9. Development of complete POP for greenhouse crops in different agro-climatic regions.
10. Identification of region-specific and techno-economic feasible cropping sequences
11. Development of effective INM, IPDM and GAP modules for protected cultivation
12. Designing and development of tools, devices and equipments for easy operations in greenhouse crops
13. Standardization of new age technologies like Hydroponics, Aeroponics, Nutrient Film Technique, Agro-voltaic systems, vertical farms *etc.*
14. Developing professional and skilled manpower
15. A mission on protected cultivation and vertical farms is needed in the country
16. Pollination management for crops under protected structures

10

CLASSIFICATION OF GREENHOUSE STRUCTURES AND SPECIFICATIONS FOR TRADITIONAL/LOW-COST GREENHOUSES

Greenhouses are framed or inflated structures covered with a transparent or translucent material in which crops are grown under partially or fully controlled environment conditions. Greenhouse structures as well as other methods of controlled environments have been developed to create favourable micro-climate in favour of better crop growth and production all the year round. Cultivation of crops under protected structures with climate controlled environment is usually intended to create off-season opportunities for the production of high valued vegetables, ornamentals, fruits *etc*. The temperature is considered as an important factor deciding the growth and development of a crop under protected environments. Simultaneously, environmental control mechanisms like cooling to mitigate excessive temperatures, light control either shading or through supplemental lights can be employed to create congenial environment for crop growth. Other parameters like carbon dioxide, relative humidity, water, nutrients and pest control can also be managed suitably with the use of modern precision faming tools.

Greenhouse structures differ a lot from simple and very cheap structures to high-tech multispan structures. The reasons for the large variety of greenhouse types in the world are mostly local conditions and the availability of building materials like timber, bamboo or steel. The overall greenhouse design is strongly influenced by the climate and latitude of a location. Regulations imposed at national or international level also influence the greenhouse design. The requirements for maximum light transmittance leads to designs with smaller and fewer structural parts, which may affect the structural stability of the greenhouse. National standards frame rules for the structural design of greenhouses and specify the loads caused by self-weight, wind, snow, crops (suspended to the structure), installations, thermal actions and maintenance.

Different types of greenhouse structures are used successfully for crop production. Although, there are advantages in each type for a particular application, in general there is no single type greenhouse, which can be considered as the best. Different types of greenhouses are designed to meet the specific needs. Many factors must be taken into consideration when constructing a greenhouse. Various elements must be looked at for greenhouse design and technology selection prior to building of protected structure. According to required functions, greenhouse styles can vary from small, stand alone structures to large, gutter connected greenhouses (multispan greenhouses). There are many designs and structures to select from have their own advantages and disadvantages. The classification of greenhouse structures is given below:

A. Types of Greenhouses Based on Shape

For classifying greenhouses on the basis of their shape, the cross section of the structure is considered as important factor. As cross section of greenhouse depicts height and width of the structure which also provides information on the overall shape of structure members. While, the longitudinal section of greenhouse is likely to be almost same for all types, hence this criterion cannot be given consideration for greenhouse classification. The details of classification based on shape of greenhouse structure are given below:

1. **Lean-to type greenhouse structure:** Lean-to greenhouses are considered traditional structures dating back to the Victorian period. A lean-to greenhouse is a type of protected structure that is built against the side of another structure (building/other structure) and typically facing South side. Therefore, it has only one sloping roof and more oftenly 3 walls, and shares a wall with another building. **Lean-to** greenhouse is usually set up on the South facing side of the building and the ridge of the roof connects to the building. This type of greenhouse is easy and cheap to heat especially in cold regions. Open side of the structure is usually oriented away from the wind and rain. The length of lean to type structures could be as long as the length of supporting building.

   These structures are cheap and easy to build using the support of building. Because of close proximity of structure to building (house), thus allows better monitoring of plants and provides easy accessibility to electricity, water *etc.* The structure makes best use of sunlight and minimizes the requirement of roof support.

However, these type of structures do have some disadvantages and these are 1) Have limited space and control over growing parameters like light, temperature and ventilation. 2) Height of the supporting building wall decides the structural design of greenhouse. 3) The structure is generally meant for single or double row plant benches with a total width of 2 to 3.5 m.

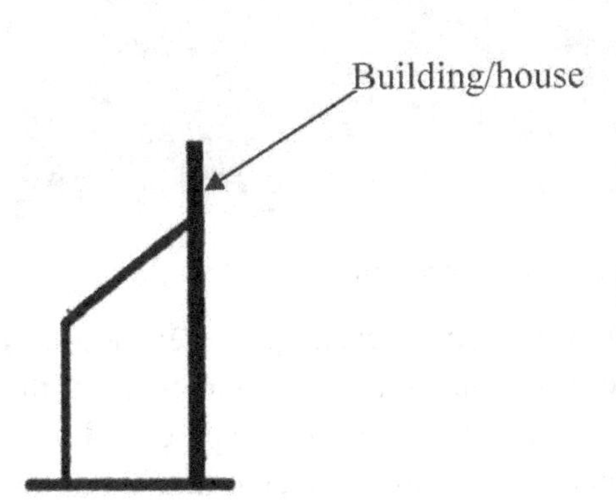

Lean-to type greenhouse structure

2. **Even span type greenhouse structure:** An even span greenhouse is constructed where pitch of both the roofs is equal in length and angle. The even span greenhouse has even pitched sides similar to that of a house. An even-span greenhouse is one of the most popular greenhouse designs because of their relative ease of construction. The efficient design provides ample growing space and is generally less costly than other greenhouse structures. Some even span greenhouses are built as an attached structure that shares a wall with another building but, unlike a lean-to greenhouse, it still retains its own symmetrical roof. Such a design is even more economical to construct. Because of its shape, the design provides better air circulation than a lean-to type structure to maintain uniform temperature inside. Connecting several single span structures makes it easy to build greenhouses called multiple span types (multispan). Generally, the span of single span type varies from 5 to 9 m, while the length is around 24 m with a height of 2.5 to 4.3 m.

   There are two basic designs of even span greenhouses namely American (also called high profile) and Dutch Venlo (also called low profile). American even-span greenhouse has one large roof per structure with overlapping panes. The Dutch even span greenhouse has two small roofs per structure. The panes on the roof of a Dutch Venlo even-span greenhouse extend from the eave to the ridge, hence there are no overlapping of panes. These greenhouses are easy to build and have even spacing, however they have overlapping panes which increases the chance of leakage.

Even span type greenhouse structure

3. **Uneven span type greenhouse structure:** An uneven span greenhouse is a type of protected structure with one roof slope longer than the other or/ the roof is not equal in width or pitch, thus called uneven span type greenhouse and the steeper angle of structure faces the South. The side that faces the South side is generally long and transparent maximizing the trap of solar radiations, while the other side facing the North is opaque for energy conservation. This is generally an adaptation of a regular greenhouse when it is situated on a hilly terrain or to take advantage of solar angles. Uneven span greenhouses are not as common anymore, as the majority of today's greenhouses are being built on flat lands. These structures work very well on the hill sides or undulating lands and the longer side of the roof allows more sunlight inside the greenhouse without any obstruction of side walls or rafters. But, these structures are more costly in comparison to a hoop house or an even span greenhouse. Long and slanting roof requires more support, which makes the roof harder to maintain over a period of time. Uneven span greenhouse located in regions closer to the equator may allow too much solar energy to penetrate inside due to its longer roof.

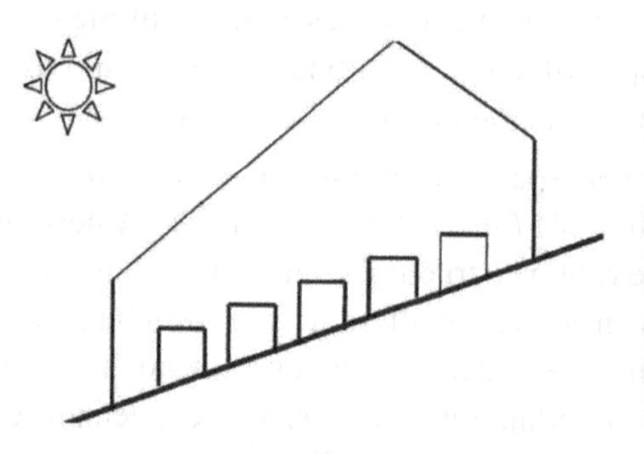

Uneven span type greenhouse structure

4. **Ridge and furrow type greenhouse structure:** A ridge and furrow greenhouse consists of a number of even span greenhouses connected along the length of the house and is also known as gutter-connected greenhouse. This design allows for an increase of space and sunlight. This type of design is composed of a number of bays or compartments running side by side along the length of the greenhouse. Typically these compartments are approximately 37 m long and 6.5 to 7.5 m wide. Ridge and furrow greenhouses may be gabled or curved arch. Gabled houses are usually suitable for heavy cladding materials like glass, fibre glass, while curved arch houses are covered with lighter materials namely polyethylene, polycarbonates. Several connected ridge and furrow greenhouses are often referred to as a 'range'. The barrel vault greenhouse consists of several quonset type greenhouses connected through gutters. An advantage is that they can be built with a ridge vent which provides air circulation throughout the structure but they are difficult to maintain. The dutch venlo greenhouse is modification of the ridge and furrow greenhouse. High winter sunlight is reached easily to grow certain crops. But, the highly ventilated windows of Dutch Venlo greenhouse can allow cold air to enter the greenhouse.

   The production area is completely open between the bays inside the greenhouse as there is no division of the structure by inner walls. The roofs of these greenhouses can either be curved or gabled. Absence of an inside wall below the gutter increases efficiency in terms of labour, automation, personal management and fuel consumption. Ridge and furrow type structures are associated with few disadvantages like improper drainage from the structure may lead to damage; shadows from gutters sometimes may prohibit sunlight and the entire production area of structure is a single space, thus may become difficult to maintain environmental parameters.

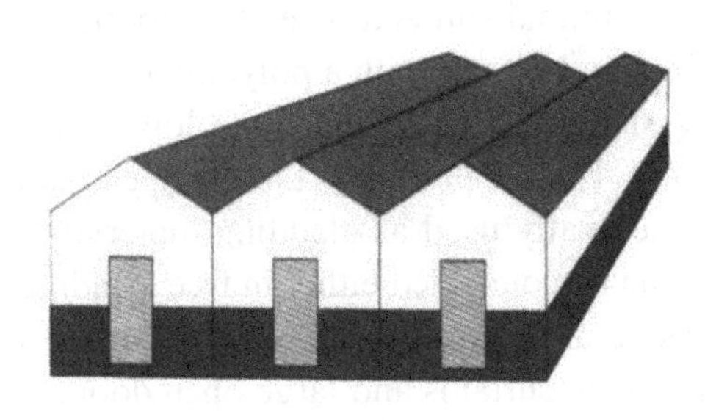

Ridge and furrow type greenhouse structure

5. **Sawtooth type greenhouse structure:**Sawtooth greenhouses are commercial structures designed with roof ventilation as the single most important feature. These structures combine optimal ventilation with special strength for withstanding different loads. The roof ventilation alone provides 25% of the total ventilation of the covered area in addition to the side ventilation. The shape of the arches allows excellent light transmission. Various standard widths and wall heights focuses on flexibility and practical use of growing space and conditions. Sawtooth vent allows a continuous airflow to reduce the inside temperature or vents can be closed to optimize the micro-climate inside the structure. Sawtooth type structures are cost-effective design and provide total protection from rain, wind and storm. The roof vent (combined with side wall vents) provides superior air transfer and hot air venting and all the vents can be closed to slow down latent heat release during winter. These structures possess high roof sawtooth openings for natural ventilation and are suitable for both cold and warm climates.

Sawtooth type greenhouse structure

6. **Quonset type greenhouse structure/Hoop-house:** This structure is characterized by a curved /dome shaped roof and military hut-style design or has a semicircular section and provides optimal sun entrance. The hoops are usually made of aluminum or PVC pipes and glazed with a polyethylene film or panels for a better insulation. The sidewalls are set up quite low which limits the storage space and headroom. This is probably the cheapest and easiest type to build. Polyethylene is generally used as cladding material in this type of greenhouse. These houses can be connected either in free standing style or arranged in an interlocking ridge and furrow. Quonset greenhouses are ventilated through roll up sidewalls/ side curtains and large open doors at both ends. The stresses of arched roof can efficiently be transferred down to the ground. These structures work well on hillsides because of the ability to maximize sun rays trapping. However, curved shape of structure limits the availability of useful interior space next to walls and the structures require more support because of the uneven structure.

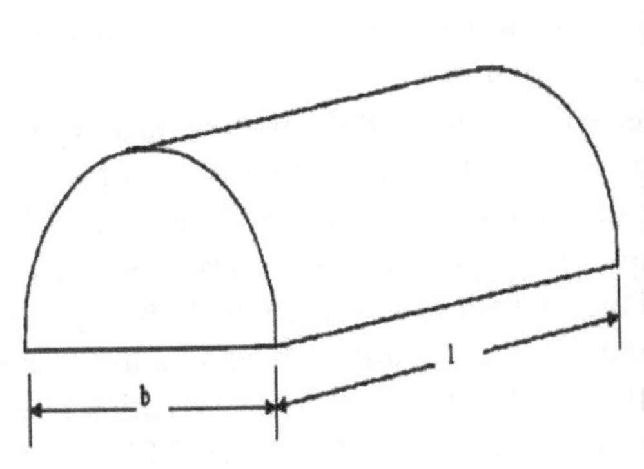

Quonset type greenhouse structure

B. Types of Greenhouses Based on Cladding Materials

Cladding (covering) materials are the most important constituent of protected structure as they have direct influence on micro-climate inside the structures. So, depending upon the type of covering materials used in the structures, the greenhouses can be classified as glass, polyhouses, rigid panel greenhouses and shade net houses.

1. **Glasshouses:** Glass is the traditional covering material for greenhouses and preferred by many growers considering greenhouse technology a long-term investment. Glass can be used on any style of greenhouse, but is commonly found in Lean-to type, even span, ridge and furrow type designs. Glass is the most expensive material to purchase and install, and the overall cost of glasshouse is substantially higher due to the increased material content of the structure.

   Greenhouse glass is available in several types including float, tempered, laminated, and frosted or hammered glass. The typical glass that is used for glasshouses is called float glass. It is a transparent glass with a high light admittance to ensure sufficient light inside the structure. With awareness among people, the use of float glass as covering material has decreased and new greenhouses are now often fitted with safety glass (*e.g.*, tempered or laminated). Tempered glass is 4 to 6 times more resistant to shattering than float glass, and even upon breakage, it splits into small square pieces making it unlikely to cause injury.

   Glasshouses have relatively long useful life span and superior light transmitting properties and less excessive relative humidity problems. Glass "breathes" (the glass laps between panes allow air to enter), while polyethylene, acrylic, and polycarbonate structured greenhouses are airtight making excessive humidity in the growing area and sometimes undesirable water drip on the plants.

2. **Polyhouses:** Flexible plastic films namely polyethylene, ethylene tetrafluoroethylene (Tefzel), polyvinyl chloride (PVC) and polyester are used as covering material in the protected structures. **Polyethylene** has always been and still is the primarily preferred film for greenhouses in most parts of world because of its low cost compared to other cladding materials. Polyethylene is also commonly known as polythene or poly/plastic film, hence greenhouse installed using such films are called as polyhouses. **Polyethylene film** is inexpensive, but is temporary and less attractive, and requires more maintenance than other plastics. It is easily destroyed by UV radiation from the sun, although film treated with UV inhibitors (UV stabilized films) will last for 2-4 years compared untreated ones. It requires fewer structural framing members for support because of its availability in wider sheets, which provides light transmission. Newly installed double layered polyethylene films reduce heat loss by 30-40% and transmit 75-87% of available light.

   Tefzel T2 film (Ethylene Tetrafluoroethylene) is another plastic coverings for greenhouse structures. A single layered Tefzel T2 film has light transmission of 95%, which is greater than that of any other covering material, while a double layer has a light transmission of 90 %.

   Polyvinyl chloride has a number of properties like excellent resistance to wear and tear, and very little effect of oxidation, thus making it desirable material for greenhouse covering. However, heat and light may break down PVC film in 2 to 3 years. PVC film has characteristic property of reducing transmission of infrared rays (long wavelength radiation), thus there are less chances of heat loss during night using PVC as a covering than using polyethylene.

   Although, the cost of **polyester film** is higher than that of polyethylene film and it is known for its durability and longer life expectancy. Other advantages include high level of light transmittance equivalent to glass and freedom from static electrical charges, which collect dust.

   Plastics as covering material for greenhouses have become popular, as they are cheap and the cost of heating is less when compared to glass greenhouses. The main disadvantage with plastic films is its short life. For example, the best quality ultraviolet (UV) stabilized film can last for four years only. Quonset design as well as gutter-connected design is suitable for using this covering material.

3. **Rigid panel greenhouses:** Rigid plastic coverings include fibre glass-reinforced plastic (FRP) rigid panel, polycarbonates, and acrylics. Light transmission through rigid plastics is very good, although it usually decreases over time with the increase of age and turn yellow due to the amount of UV

radiation. The large sheets rigid plastics are much lighter than glass and require fewer support bars for their attachment to the greenhouse frames. However, these rigid panels are not so easy to install on curved roofs.

Fibre glass reinforced plastic (FRP) rigid panel has been in use as greenhouse coverings since 1950s but its popularity has declined in recent years. Fibre glass is available in flat and corrugated configurations. Corrugated panels are commonly used for greenhouse roofs, as its corrugated shape lends strength and rigidity to the panels. Flat panels are usually used for sidewalls, windows and vents. Although, FRP panels are classified as a rigid plastic, but they are flexible enough to be bent in a curve to fit the framework of a Quonset type or arch type greenhouse.

Polycarbonate is one of the most widely used structured sheet materials in greenhouses today. Polycarbonate sheets have a higher percentage of direct radiation than diffused radiation, when compared with polyethylene film. It is similar to acrylic in heat retention properties and allows about 90% of the light transmission of glass. Polycarbonate has slightly less light transmission compared to acrylic, but is considerably stronger and impact resistant, more flexible, and only flammable when an active flame comes in contact with the material. Although, the initial cost of polycarbonate is high, but 10 to 15 years life span can be expected (depending on the manufacturer). Polycarbonate has high impact strength about 200 times that of glass.

Acrylic has been used for many years and is considered to be the most suitable rigid transparent plastic for greenhouse glazing. Acrylic is flame retardant and UV stabilized, and has excellent clarity and light transmission, high impact resistance textured surface, which diffuses light thus preventing condensation drip. However, these panels tend to collect dust as well as to harbor algae, which results in darkening of the panels and subsequent reduction in the light transmission.

4. **Shading nets:** There are a great number of types and varieties of plants growing naturally in the most diverse climate conditions that have been transferred by modern agriculture from their natural habitats to controlled crop conditions. Therefore, conditions similar to the natural ones must be created for each type and variety of plant. Each cultivated plant must be given a specific type of shade required for the diverse phases of its development. The shading nets fulfill the task of giving appropriate micro-climate conditions to the plants. Shade nettings are designed to protect the crops and plants from UV radiation, but they also provide protection from climate conditions such as temperature variation, intensive rain and winds. Better growth conditions can be achieved for the crop due to the controlled micro-climate conditions 'created' in the covered area with shade netting,

which results in higher crop yields. All nettings are UV stabilized to fulfill expected lifetime at the area of exposure and have characteristic of high tear resistance, low weight for easy and quick installation.

A wide range of shading nets is available in the market which is defined on the basis of the percentage of shade they deliver to the plants growing under them. Shade nets are available in different shade percentages or shade factor *i.e.* 15%, 35%, 40%, 50% 75% and 90% (for example 35% shade factor means the shade net will cut 35% of light intensity and would allow only 65% of light intensity to pass through the net).

Screen houses: Screen houses are the structures covered with insect screening material instead of plastic or glass. They provide environmental modification and protection from severe weather conditions as well as exclusion of pests. They are often used to get some of the benefits of greenhouses in hot or tropical climate.

Crop top structures: A crop top is a structure with a roof but doesn't have walls. The roof covering may be made up of plastic film or glass, shade net or insect covering. These structures provide some modification of the growing environment such as protection of the crop from rain or reduction of light levels.

Glasshouse

Polyhouse

Rigid panel greenhouse

Shade net house

C. Types of Greenhouse Based on Utility

Greenhouses are generally designed to provide a controlled environment for plant production with sufficient sunlight, temperature and humidity. As temperature is one of the important growing factors deciding the success in crop production under protected environment, so desirable level of temperature needs to be maintained for better growth and development of crop plants. Specific temperature requirements of a particular crop species is also linked to geographical location of greenhouses. Greenhouses located in temperate needs heating for year round production of crops and similarly greenhouses in tropical and sub-tropical regions need cooling to maintain desirable temperature conditions. For maintaining temperature either through heating or cooling inside greenhouse depending upon locations, energy is required to make effective heating or cooling. Therefore, protected structures can be classified as greenhouses for active cooling and active heating depending on the purpose of cooling and heating.

1. **Greenhouses for active cooling:** Under tropical and subtropical conditions, it is important to maintain temperature inside the greenhouses at desirable level for successful cultivation of crops. Therefore, greenhouses need to be designed location specifically to allow better natural air inflow for reducing temperature inside the greenhouses. Ventilation not only reduces inside temperature during sunny days but also help to supply carbon dioxide, a vital input for photosynthesis. Another advantage of ventilation is to remove warm, moist air and replace it with dry air. High humidity is objectionable since it causes moisture condensation on cool surfaces and tends to increase the occurrence of diseases. However, sometimes it becomes very difficult to create suitable environment inside greenhouses during peak summer season, therefore cooling system involving energy has to be installed. In such cases, evaporative cooling can be achieved via fan and pad system or by fogging system.
2. **Greenhouses for active heating:** Greenhouses under temperate regions need heating for year round crop production. A good heating system is one of the most important steps for successful plant production. Any heating system that provides uniform temperature control without releasing any harmful material to the plants is acceptable. Energy sources like natural gas, LP gas, fuel oil, wood and electricity can suitably be used for the purpose of heating, but due consideration has to be given for their cost and availability. Convenience, investment and operating costs are also the other considerations need more planning for economic gain from greenhouses. Greenhouse heater requirements depend upon the amount of heat loss from the structure. Heat loss from a greenhouse usually occurs by all three modes of heat transfer namely

conduction, convection and radiation. Many types of heat exchange occur simultaneously and the heat demand of a greenhouse is normally calculated by combining all three losses as a coefficient in a heat loss equation.

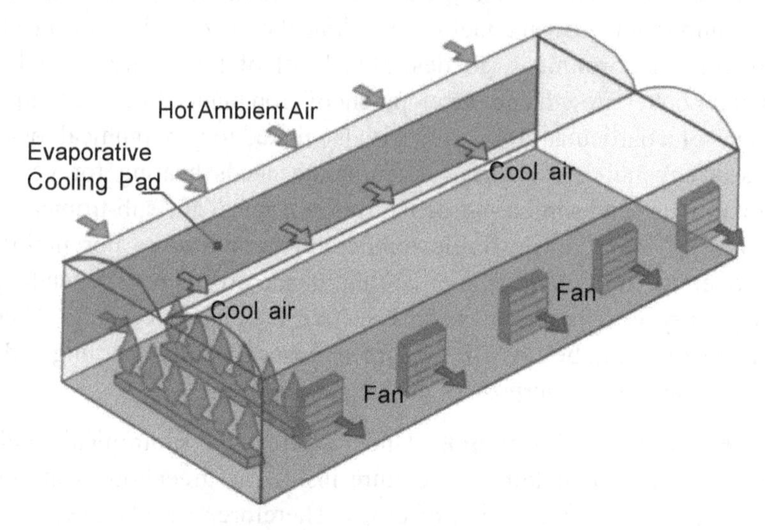

Greenhouses for active cooling

Greenhouses for active heating

D. Types of Greenhouses Based on Construction

The construction of greenhouses is largely dependent on the material used for erection of structures and cladding material. Selection on the type of members is determined by the span height of protected structures, so stronger material is required to construct a greenhouse with higher span. Simultaneously, more number of structural members are needed to make the greenhouse stronger depending upon the design. Hence, greenhouses can be grouped in three broad categories namely wooden framed, pipe framed and truss framed.

1. **Wooden framed greenhouses:** Wood is the most commonly used material with home or local contractor for erecting greenhouses and to a limited extent by some commercial greenhouse manufacturers for the gothic shaped structure and for glass supporting bars. Wooden framed structures are generally meant to build the greenhouses with span less than 6 m. It is mandatory to use the most decay resistant wood species or treated lumber under high moisture environment to achieve reasonably long and desired life. Wood treatment can be done with one of the water borne salt type preservatives such as Chromated Copper Arsenate, Ammonical Copper Arsenite or Fluor Chrome Arsenate Phenol. Treatment of wood with such preservatives may increase the longevity for 15-20 years or more so as the life of greenhouse structure.

   Many of the designs are based upon selection of structural quality, which is the highest grade normally used in construction. Some of the designs further specify that any defects in the lumber may be placed out of the high stress areas. With the rigid framed designs, this high stress area is the stud and rafter closest to the eave joint.

Wooden framed structures

2. **Pipe framed greenhouses:** The choice of construction of pipe framed greenhouses often favours low initial investment and relatively longer life. Galvanized pipes and UV stabilized low density polyethylene (LDPE) film are common options of greenhouse fabricators. Pipe framed greenhouses are well suited to accommodate a span of 12 m and usually have one of the members (rafter, chord, strut) missing from the protected structure. As these three members constitute technically 'truss', which is not used in pipe framed structures and the components of pipe are attached to sash bars for better support but they are not connected together.

In most of the cases, GI pipes are used in greenhouses ranging from small houses to huge multi-bay commercial greenhouses. Galvanized pipes generally have the rational cost effectiveness in the market. Compared with other typical steel pipe coatings, such as specialized painting and powder coating, galvanization is much more labour-intensive, resulting in a higher initial cost for contractors. Besides, due to its durability and anti-corrosive properties, galvanized pipe can be recycled and reused, which save money during the post maintenance work to some degree.

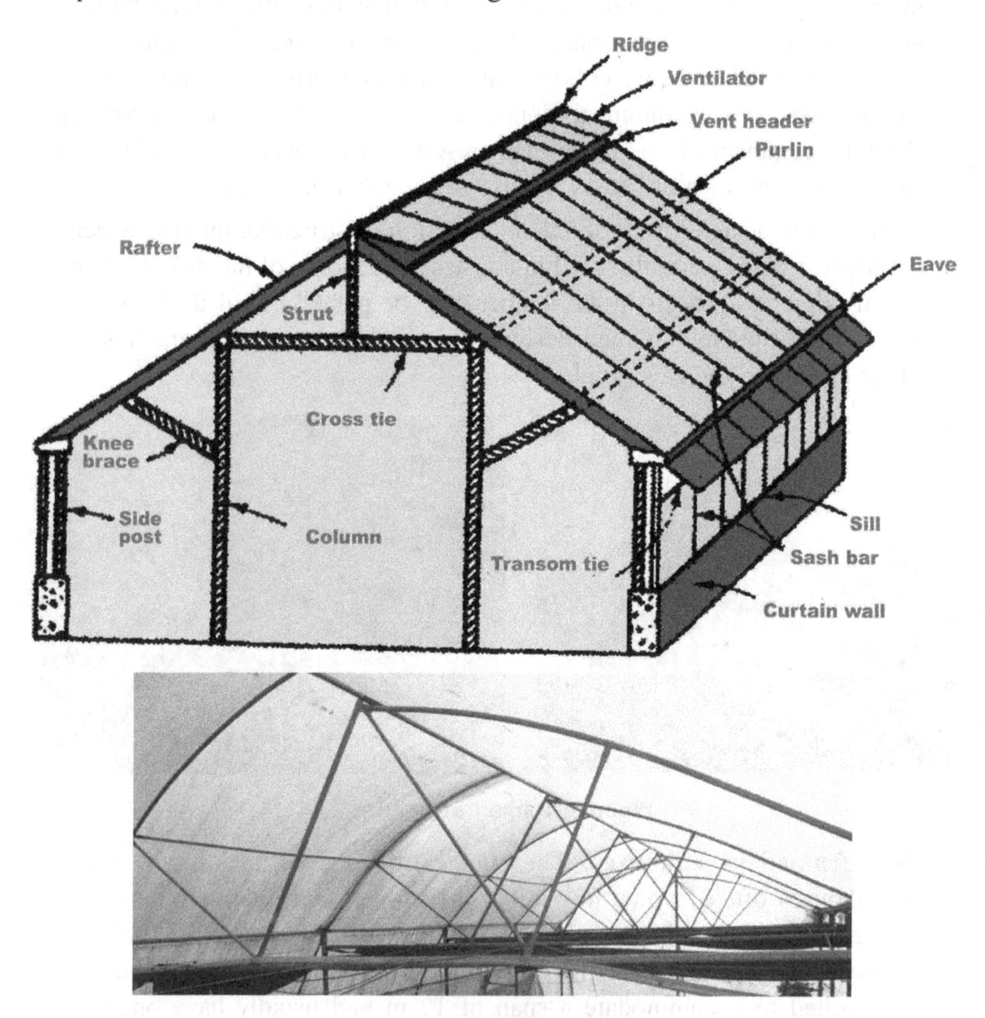

Pipe framed structures

3. **Truss framed greenhouses:** Truss framed structures are suitably constructed to accommodate a span of 15 m or more than that. Three members of greenhouse namely rafter, chord and strut jointed together to form truss in these type of greenhouses. Both struts and chords serve as support members

of greenhouse, first one (strut) works under compression while later one performs under tension. Purlins of structure running through the length of greenhouse are joined to truss for better strength of the structure.

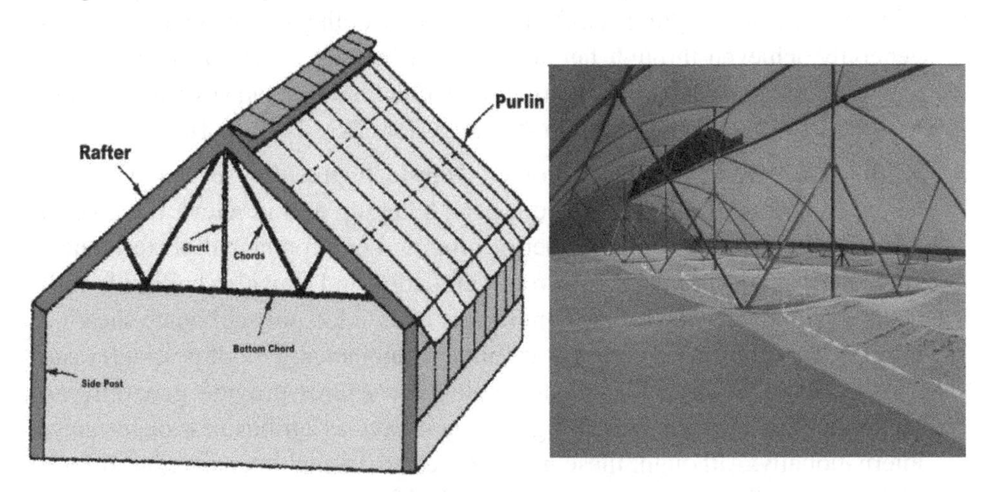

Truss framed structures

E. Types of Greenhouses Based on Structure Cost

1. **Low cost/ Low-tech greenhouse:** Low cost greenhouse are often constructed with locally available materials such as bamboo, timber *etc.* and have height less than 3 m. These greenhouses lack vertical walls and have poor ventilation. Very little or no automation is followed in such structures. Though, these structures provide basic advantage over open filed cultivation, however crop potential is still limited by the growing environment and the crop management is relatively difficult. Such protected structures generally result in a sub-optimal growing environment which restricts yield of crop and have very little role to reduce population of insect-pests and diseases. Low cost greenhouses have significant production and environmental limitations, but offer a cost-effective approach of cultivating crops out of the season. These greenhouses generally work on passive cooling (natural ventilation) for maintaining temperature and humidity to some extent. Cost of such structures is around Rs. 300 to 600 per m^2.
2. **Medium cost/Medium-Tech greenhouse:** Medium cost greenhouses are typically characterized by vertical walls more the 2 m but less than 4 m tall and a total height usually less than 5.5 m. These greenhouses may have roof or side wall ventilation or both. These greenhouses are usually glazed with either single or double layer plastic films or glass and use varying degree of automation. Medium-tech greenhouses offer a compromise between cost and productivity, and represent a reasonable economic and environmental basis

for the greenhouse industry. Production in such greenhouses can be more efficient than open field production. There is a greater opportunity to non-chemical pest and disease management strategies but overall the full potential of greenhouse horticulture is difficult to attain. Cooling in such greenhouses is generally achieved through fan and pad and fogging. The cost of medium-tech greenhouse with natural ventilation without fan and pad system is around Rs. 800 to 1100/m^2 and with fan & pad system (Rs. 1400 to 1600/m^2).

3. **High cost/ High-tech greenhouse:** High-tech greenhouses have a wall height of at least 4 m, with the roof peak being up to 8 m above the ground level. These structures offer superior crop and environmental performance. High technology structures have roof ventilation and may also have side wall vents. Cladding may be plastic film (single or double), polycarbonate sheeting or glass. Environmental controls are almost automated. Use of pesticides can be significantly reduced. High technology structures provide generally an impressive sight and are increasingly being viewed as agribusiness opportunity internationally. Although, these greenhouses are capital intensive, but offer a highly productive, environmentally sustainable opportunity for an advanced fresh produce industry. Cost of high tech greenhouses with fully automatic system is around Rs. 3000 to 4000/m^2.

F. Types of Greenhouses Based on Environmental Control

1. **Greenhouse with natural ventilation:** Greenhouses in which environmental parameters like temperature, humidity *etc.* are regulated without involving external energy through natural ventilation, are termed as naturally ventilated greenhouses. As the name suggests, these greenhouses do not have any specific environmental control system except for the provision of adequate ventilation and fogger system to prevent basically the damage from weather aberrations and other natural agents.
2. **Greenhouse with active ventilation:** These greenhouses involve the use of external energy for regulating environmental parameters inside the structures. Such structures are fitted with active summer or winter cooling system depending upon the locations along with other accessories to extend the growing season or permits off-season production by way of controlling light, temperature, humidity, carbon dioxide level *etc.*

G Types of Greenhouses Based on Application of System

1. **Greenhouse for crop production/cultivation:** These greenhouses are primarily meant for cultivation of different horticultural crops like vegetables, flowers and fruits. Cultivation of different crops under greenhouses requires

specific temperatures and relative humidity during certain phenological stages. Therefore, greenhouses meant for crop production must have sufficient control over different environmental parameters for overall growth and development of crops especially for off-season production.

2. **Greenhouse for crop drying:** Drying under sun using the solar radiations for food preservation is practiced since ancient times. Sun drying is seasonal, intermittent, slow and unhygienic; therefore a suitable alternative needs to be sought. The moisture content and temperature at which food product is to be dried is always fixed. Proper maintenance of the said factors is possible only in controlled mechanical drying. The energy demand of conventional mechanical dryers is met by electricity, fossil fuels, firewood and these sources are costly and becoming scarce. So, solar energy can be an alternative source for drying of food through solar dryers. So, the advanced method of drying *i.e.* greenhouse drying is being adopted to overcome the limitations of traditional (open sun) method. The product is generally placed in trays receiving the solar radiations through cladding material and moisture is removed by natural or forced convection/ air flow. A maximum drying efficiency of around 65% can be achieved through greenhouse drying.
3. **Greenhouse as solar energy collector:** Solar greenhouses are designed not only to collect solar energy during sunny days but also to store heat for use at night or during periods when it is cloudy. They can either stand alone or be attached to houses or farm buildings. A solar greenhouse may be an underground pit, a shed-type structure, or a Quonset hut. Large scale producers use freestanding solar greenhouses, while attached structures are primarily used by home scale growers. Passive solar greenhouses are often good choices for small growers, because they are a cost-efficient for farmers to extend the growing season. In colder climates or in areas with long periods of cloudy weather, solar heating may need to be supplemented with a gas or electric heating system to protect plants against extreme cold. Active solar greenhouses use supplemental energy to move solar heated air or water from storage or collection areas to other regions of the greenhouse.

Solar greenhouses differ from conventional greenhouses in the following four ways:

- Have glazing oriented to receive maximum solar heat during the winter
- Use heat storing materials to retain solar heat
- Have large amounts of insulation where there is little or no direct sunlight
- Use glazing material and glazing installation methods that minimize heat loss
- Rely primarily on natural ventilation for summer cooling

The two most critical factors affecting the amount of solar heat absorbed by a greenhouse are:

- The position or location of the greenhouse in relation to the sun
- The type of glazing material used.

Specifications for Traditional and Low Cost Greenhouses

Wood/ Bamboo based naturally greenhouse comes under traditional and low cost greenhouses. Wood/ Bamboo based naturally ventilated greenhouse (also called rain shelter in heavy rain fall areas) are made of Casurina, Nilgiri, Bamboo *etc.* The poles made of these materials should preferably be treated either with turpentine or coal tar at one end that will be placed in the foundation pits. Aluminum profiles with spring or wooden battens may be used for fixing the UV stabilized poly ethylene films (cladding material). The wooden/ bamboo poles and other supporting materials used in the greenhouse structure need to be strong enough to withstand different types of load like wind load, crop load *etc.* To withstand such loads, the recommended specifications of outer diameter of wooden/ bamboo poles lies between 8 to 10 cm for main poles and 6 to 8 cm for purlins and trusses. Preferably, all parts of the structure are fitted with the help of nails or nut-bolts.

The technical specifications for a wooden or bamboo structure greenhouse are given below:

Technical specification for Wooden/ Bamboo based greenhouse of 100 m^2- 250 m^2 size:

Sr. No.	Item	Description/ Specifications
1.	Width of structure	At least 35% of the desired length of greenhouse
2.	Ridge height	3.5 m to 4 m
3.	Ridge Vent	80 cm to 90 cm opening fixed with 40 mesh nylon insect screen/ insect proof net
4.	Gutter height	2.25 m to 2.75 m from floor area
5.	Gutter slope	The gutter slope should be at least 2%
6.	Structural design	The structural design should be sound enough to withstand minimum wind speed of 130 km/hr and minimum load of 20 kg/m^2. There should be provision for opening one portion at either side for power tiller entry to perform intercultural operations.

[Table Contd.

Contd. Table]

Sr. No.	Item	Description/ Specifications
7.	Structure	Wooden posts should have dimension 8 to 10 cm diameter for central post, side post and Gutter post/ tie beam *etc*. And diameter 6 to 8 cm for post plate, supporting post, trusses/ members/sticks/others structural members for joining each other properly.
1.	Treatment of poles	The post must be treated with different type of preservatives to protect it from termites/fungal attacks. The recommended preservatives are Coal Tar Creosote, Copper Zinc Naphthenates and Abietates, Boric Acid and Borax, Copper Chrome Arsenic (CCA) Composition, Acid Curpric Chromate Composition and Copper Chrome Boric Composition.
1.	Fasteners	All nuts & bolts, nails, Aluminum/ MS strip of 2 cm width must be of high tensile strength and galvanized and if required, there should be provision of PP/ Coir & Jute ropes for anchoring the structure.
8.	Entrance room & Door	Two entrance door of size 2.5 m x 1.5 m must be provided as per the requirements and covered with 200 micron UV stabilized transparent plastic film.
9.	Cladding material	UV stabilized 200 micron transparent plastic films conforming to Indian Standards (IS15827:2009), multilayered, anti drip, anti fog, anti sulphur, diffused, clear and having minimum 85% level of light transmittance.
10.	Fixing of cladding materials	All ends/ joints of plastic film need be fixed with two way Aluminum profiles with suitable locking arrangement along with curtain top.
11.	Spring Insert	Zigzag high carbon steel with spring action wire of 2-3 mm diameter must be inserted to fix shade net into Aluminum Profile.
12.	Curtains and insect screen	Roll up UV stabilized 200 micron transparent plastic film as curtains need be provided up to recommended gutter height on all sides having provision for manual opening and closing of curtains. 40 mesh nylon insect proof nets (UV stabilized) of equivalent size need to be fixed inside the curtains. Anti flapping strips are also suggested to ensure smooth functioning of the curtain.
13.	Shade net	UV stabilized 50% shading net with manually operated expanding and retracting mechanism. Size of net should be equal to floor area of the greenhouse.
14.	Footpath	1 m wide and 10 cm thick footpaths should be provided in the centre (Length x width) and made of cement concrete 1:2:4.

[Table Contd.

Contd. Table]

Sr. No.	Item	Description/ Specifications
After sales services		
15.	Warranty	The firm to provide warranty frees maintenance for one year from the date of installation.
16.	Testing	All plastic materials used in the greenhouse should be tested by the CIPET for quality assurance (If necessary).
17.	Training	Free training for operation, maintenance and production for one year.

11

AGRIVOLTAIC SYSTEM: CONCEPT AND FEATURES

Among all the primary, domestic and clean or inexhaustible renewable energy resources, solar thermal energy is the most abundant one and is available in both direct as well as indirect forms. Solar energy conversion using photovoltaic systems have led to real estate speculation with concomitant price escalation and dire social consequences. Improvements in productivity using mechanization through fossil fuel revolution are now marginal thus leading to demand for carbon free electricity.

Agrivoltaic

Electricity is generated from Photo voltaic (PV) panels mounted in designed spacing and height, so that limited shading allows for productive agriculture on land below. In Japan, concept of agrivoltaic is also known as "Solar Sharing".

Agrivoltaic System

Two main factors for agrivoltaics

1. With adequately designed spacing PV installations permit considerable sunlight to reach ground level.
2. Plant species exhibit different tolerance level to shading and different levels of light saturation point.
    - Plant growth increases with amount of incident light reaching the light saturation point, and further increase in light intensity does not lead to further growth.
    - The excess and unutilized light can be detrimental to plant growth and quality.
    - A well designed agro-voltaic power plant could therefore produce similar agricultural yield compared to open field cultivation and generate electricity as additional output.

Features

- Electricity generation where it is not accessible via grid.
- Photovoltaic energy generation has great potential to reduce greenhouse gas emission.
- Provides additional income and employment opportunities in rural India.
- Easy light management for shade loving crops.
- Allowing normal crop growth in tropics where season is abnormal and dull with high temperature and intense sunlight.
- Village electrification.
- In case of total crop failure during drought or aberrant weather, the income from PV system is ensured.

Photovoltaic Greenhouse

- Cohering the capability of enhanced productivity and resource use efficiency of greenhouse with the renewable energy production; photovoltaic greenhouse (PVG) shows a credible solution.
- Energy is largest overhead cost in production of agricultural greenhouses in temperate regions.
- Socio-economic and environmental efficiency can be higher in combined photovoltaic greenhouse than in the solar or greenhouse system independently.
- Use of semi-transparent (photo selective) PV modules seems to be promising approach for PV greenhouse models.

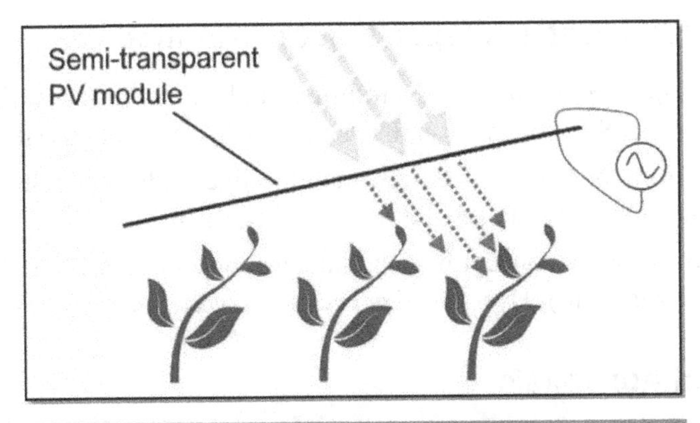

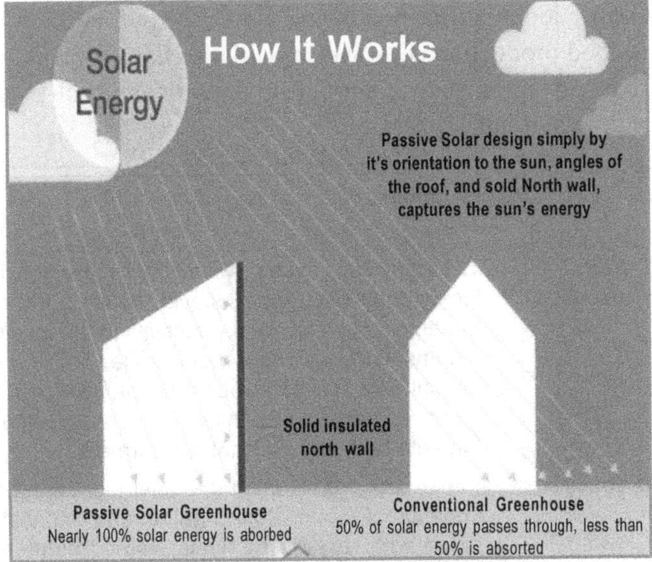

Working of solar greenhouse

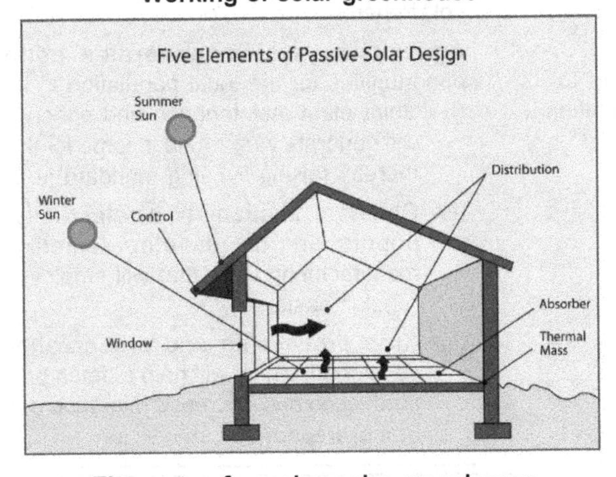

Elements of passive solar greenhouse

Types of cells used in preparation of solar modules

1st Generation: Silicon based modules such as monocrystalline (c-Si), polycrystalline (p-Si), ribbon crystalline silicon (r-Si), amorphous silicon (a-Si).

2nd Generation: Non-silicon based modules like cadmium telluride (CdTe), copperindium (gallium) and selenide (CIS or CIGS).

3rd Generation: New concept devices such as concentrated PV (CPV), organic cells (single or tandem), Dye-sensitized cells (Gratzel), Perovskite cells.

Benefits of Agrivoltaic

Agrivoltaic allows double utilization of the same piece of land for generation of electricity and food production. The System will also have the following features and benefits that will significantly contribute to the economic and general progress of the rural people particularly farmers.

Sr. No.	Features	Benefits
1.	Growing crops during dry season	The Agrivoltaic system will allow for the growth of shade loving crops like ginger, stevia, gourds, and culinary herbs thereby increasing the productivity of the land during the dry months while making it more fertile. These crops will allow agricultural activity throughout the year and increase the number of crop cycles to two or more and also open up new markets and revenues for the farmers.
2.	Water efficiency	Water used for cleaning the solar plant can be recycled for irrigation of crops. Drip irrigation and rain water harvesting along with reverse osmosis water treatment will ensure optimum usage of water. Water will also be pumped using solar energy.
3.	Source of employment to local population	Agrivoltaic system generates better employment opportunities for the local population in three areas (i) Solar plant maintenance and agricultural activity. This will generate year round income for the local population thereby raising livening standard in the region. (ii) Constant availability of electricity will help rural population to develop small business and manufacturing units that will employ rural people on a regular basis. (iii) Crop processing and preservation units can be developed, which will help to fetch better return of farm production and decrease migration of rural youth to the urban areas.

Major limitations of Agrivoltaic

- Renewable energy generation requires high and large scale initial investments.
- In the developing countries like India, adoption of the system at larger scale is possible through the support of conspicuous public subsidies.
- This system may occupy large abandoned agricultural area.
- Lack of proper characterization of micro-climate inside photovoltaic greenhouses with high level of shading.

Future prospects

- Application of photovoltaic system in soil solarization.
- Solar renewable energy offers complimentary solution to many wine growing processes through concept of solar winery.
- New efficient and cost effective ways of solar energy storage have to be developed.
- Suitable species and appropriate cultivation technique of cultivation must be defined in order to counterbalance radiation reduction under photovoltaic greenhouses.
- The concept of agrivoltaics is at early stage of development and the performance needs to be monitored closely to evaluate in context of sustainability.

12

PLANT RESPONSE UNDER GREENHOUSE PRODUCTION SYSTEM

The performance of plant species is largely dependent on its genetic makeup but suitably controlled by different set of environmental conditions. A plant species/variety which performs very well in a given set of environment may have differential response in another environment. This clearly depicts the importance of micro-climatic factors like light, temperature, relative humidity, carbon dioxide which affect overall growth and development of a plant. Therefore, knowledge of interaction between plants and their surrounding environment is very much helpful to understand plant requirements and develop better management strategies for greenhouse production system.

Light

Light energy is used by plants in the process known as photosynthesis to manufacture carbohydrates and then these carbohydrate molecules are used by plants to produce other important compounds. Respiration is the process in which the trapped solar energy is later released. Solar radiation reaching the earth surface can be viewed as UV-radiation (ultraviolet), visible light PAR (violet, blue, green, yellow, orange and red), NIR (near infrared) and FIR (far infrared) [**Fig. 12.1a, b**]. The light level in the photosynthetically active radiation (PAR, 400-700 nm) is mostly decisive for photosynthesis following the sensitivity curve of the crop (the red part is the most effective; the green part the less effective). The spectral criteria like the relation of red/far red and the quantity of blue light are very important for crop development. The blue light not only contributes to the photosynthesis but also to the photomorphogenesis. The rate of photosynthesis peaks around 450 nm (blue light) and 650 nm (red light) in most of the plants (**Fig. 12.2**). The spectral distribution of radiations characterized by the wavelength is given in **Table 12.1**.

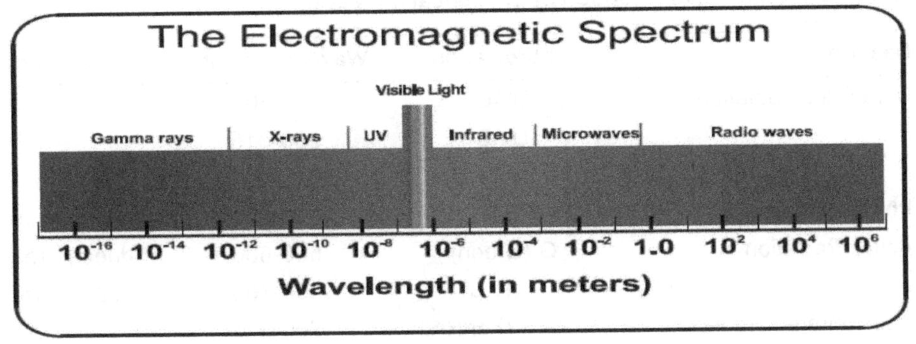

Fig. 12.1a

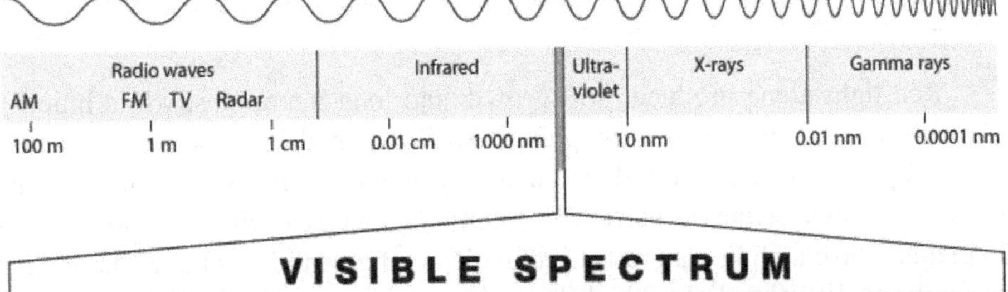

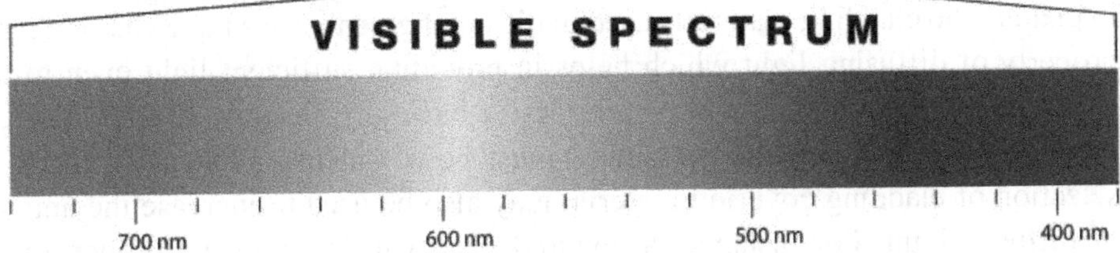

Fig. 12.1b

Fig. 12.1a, b: Partitioning of solar radiation

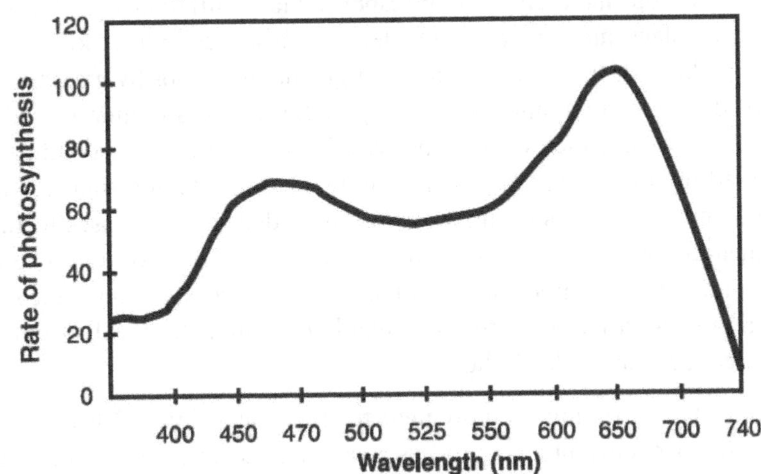

Fig. 12.2: Relation of solar spectrum and rate of pphotosynthesis in plant species

Coustesy: Badgery-Parker, 1999.

Table 12.1: Spectral distribution of the optical radiation

Description	Abbreviation	Wavelength (nm)	Remarks
Ultraviolet Radiation	UV-C	< 280	
	UV-B	280-315	
	UV-A	315-400	
Photosynthetically Active Radiation	PAR B (Blue)	400-500	<300 nm doesn't reach the earth
	G (Green)	500-600	
	R (Red)	600-700	
Near Infrared Radiation	NIR FR (Far Red)	700-800	
	NIR	800-3000	
Far Infrared Radiation	FIR	3000-100000	

Red light alone produces soft growth and long internodes, while blue light alone results in short and hard plants. So, a balance of light across the PAR range is preferable and the reduced intensity across the full spectrum is assumed to be better than reducing particular colour wavelengths while providing shades to plants. The cladding materials used in different greenhouses have characteristic property of diffusing light which helps in providing sufficient light even to the lower parts of the plant canopy and doesn't have any shadow effect. The light coming inside the protected structure must be of sufficient intensity (lux), so selection of cladding/covering material may also be used to increase the amount of diffused light. For instance, a textured glass can increase the proportion of diffuse light without significantly reducing the total level of light transmitted.

Different crops have differential response under different light levels, but in general many plants may tolerate light intensities as low as 25 to 30 klux. Generally, an optimum level of light is maintained for greenhouse crops by maximizing light intensity during winter and minimizing radiation in summer to reduce the accompanying heat, which would otherwise force plants to stress condition. Light is increased by minimizing objects above the plants including frames, pipes and other equipment/ or artificial light may be provided in some cases to fulfill the requirement of plant. The level of radiation entering a greenhouse can be reduced during summer by the white washing or suspending shade net above the structure. There are also a variety of screens available to reduce the level of radiation in order to reduce heat and/or light.

Plants have an optimal requirement for light intensity which represents the point at which photosynthesis process is maximized to achieve optimum plant growth. For example, a light level of 4 klux is just enough for the rate of photosynthesis to equal the rate of respiration in chrysanthemum, which is termed as the light compensation point. At this point, there is no net growth, but the plant can survive. The point where an increase in light intensity will not increase photosynthesis any more is called light saturation. In many crops, an upper leaf would be saturated at around 32 klux. However, because of shadowing of lower leaves, light levels of around 100 klux might be necessary for whole plant to become light saturated. In a greenhouse, the intensity of light can range from as much as 130 klux on a clear summer day to less than 3 klux on overcast days.

It is important to distinguish between diffuse and direct radiation in respect of PAR radiation. Direct radiation refers to the radiation, which reaches the earth directly from the sun without being reflected. However, most of the sun radiations is scattered by small water droplets in the atmosphere on cloudy days and reaches the earth as diffuse radiation. When the radiation is direct, there are big differences between the locations in the greenhouse with and without shadow (40% compared to 85%). The transmittance for diffuse light is in general 10% lower than the transmittance for direct light but it reaches the lower parts of high plants better, which is positive for plant growth. The transmittances for diffuse and direct (perpendicular) light of some cladding materials are given for the wavelength band of 400-700 nm (PAR) in **Table 12.2.**

Table 12.2: Transmittance of greenhouse cladding materials for direct and diffuse light

Cladding Material	Thickness	Transmittance for direct light (%)	Transmittance fordiffuse light (%)
Glass	4 mm	89-91	82
PE (Polyethylene) film	200 μm	89-91	81
EVA (Ethylenevinylacetate) film	180 μm	90-91	82
PVC (Polyvinylchloride) film	200 μm	87-91	–
ETFE (Ethylene tetra fluoroethylene) film	100 μm	93-95	88
PVDF (Polyvinylidene fluoride) film	100 μm	93-94	85
PC (Polycarbonate) sheet	12 mm	80	61
PMMA (Poly methylmethacrylate) sheet	16 mm	89	76
PC (Polycarbonate) zigzag sheet (double)	25 mm	88	79

Ultraviolet (UV) radiation can be separated as UV-A (315-400 nm), UV-B radiation (300-315 nm) and UV-C radiation (< 280 nm) and has a high energy content. UV-A and UV-B are responsible for degradation of plastics and therefore it is necessary to protect greenhouse plastic films for UV-degradation (ageing) by adding UV-stabilizers to the polymer. In combination with blue light, UV-A plays a role in some morphological processes like stem stretching and leaf development. UV-B can influence the colouring of flowers by developing pigments like flavonoids.

Near infrared (NIR) is that part of solar spectrum which is hardly used by the plants for photosynthesis. It is mostly substituted into sensible and latent heat in the greenhouses. This could be an advantage in a country with a colder climate. However, it is not good for greenhouses located in warmer countries like in the tropics and subtropics as NIR radiation leads to additional unwanted heating in greenhouses. So, special plastic films have been developed blocking the NIR to limit heating process in tropical countries and may have some cooling effect.

Far infrared radiation (FIR) is not the part of direct sun radiation, but it is heat radiation transmitted by each 'heat body' in and on the greenhouses has very important role to execute greenhouse effect. Glass and other greenhouse covering materials are transmitting this long wave radiation in a certain extent. To avoid strong radiative heat losses during cold clear nights a covering material should reflect FIR and not transmit it. In subtropical regions, there is a need for thermic films, which prevent the transmission of FIR during cold nights and thereby create an insulating effect.

Temperature

Development and flowering in plants relates to both air and root zone temperatures and management of temperature is an important tool for maintaining proper crop growth .

Air temperature

The optimum temperature is determined by the processes involved in the utilization of photosynthetic assimilates i.e. distribution of dry matter to shoots, leaves, roots and fruits. Average temperature over a period of time (one or several days) is more important than the day/night temperature differences for the management of crop growth, which is also referred to as the 24 hour average temperature or 24 hour mean temperature. Various greenhouse crops show a very close relationship between growth, yield and the 24 hour mean temperature. With the goal of directing growth and maintaining optimum plant balance for sustained high yield production, the 24 hour mean temperature can be manipulated to direct the plant to be more generative in growth or more vegetative in growth. Optimum photosynthesis occurs between 21 to 22°C, this temperature serves as the target for managing temperatures during the day during the period of photosynthesis. For example, optimum temperature for vegetative growth of greenhouse peppers lies between 21 to 23°C with an optimum temperature of about 21°C for yield.

Fruit set, however, is determined by the 24-hour mean temperature and the difference in day-night temperatures, with the optimum night temperature for flowering and fruit setting at 16 to 18°C. Target 24 hour mean temperatures for the main greenhouse vegetable crops (cucumber, tomato, pepper) can vary from crop to crop with differences even between cultivars of the same crop. The optimum 24 hour mean temperature for vegetable crops ranges from 21 to 23°C depending on light intensity. The general management strategy for directing the growth of a crop is to raise the 24 hour average temperature to push the plants in a generative direction and to lower the 24 hour average temperature to encourage vegetative growth. Adjustments to the 24 hour mean temperature are made usually within 1 to 1.5°C with careful attention paid to the crop response. One assumption that is made while using air temperature as the guide to direct plant growth is that it represents the actual plant temperature. The role of temperature in the optimization of plant performance and yield is ultimately based on the temperature of the plants. Plant temperatures are usually within a degree of air temperature, however during the high light periods of the year, temperature of plant tissues exposed to high light may reach 10 to 12°C higher than air temperatures. Therefore, considering this fact one must use strategies such as shading and evaporative cooling to reduce overheating of the plant tissues. Infrared thermometers are useful for determining actual leaf temperature.

Precision heating of specific areas within the crop canopy add another dimension of air temperature control beyond maintaining optimum temperatures in the entire greenhouse. Using heating pipes that can be raised and lowered, heat can be applied close to flowers and developing fruit to provide optimum temperatures for maximum development in spite of the day-night temperature fluctuations required to signal the plant to produce more flowers. The rate of fruit development can be enhanced with little effect on overall plant development and flower set. Precise application of heat in this manner can avoid the problem of low temperatures to the flowers and fruit which are known to disturb flowering and fruit set.

The functioning of bell pepper flowers are affected below 14°C , the number of pollen grains per flower are reduced and fruit set under low night temperatures leads to the production of deformed fruits. Problems with low night temperatures can be sporadic in the greenhouse during the cold winter months and can occur even if the environmental control system is apparently meeting and maintaining the set optimum temperature targets. There can be a number of reasons for this, but the primary reasons are 1) lags in response time between the system's detection of the heating set point temperature and when the operation of the system is able to provide the required heat throughout the greenhouse and 2) specific temperature variations in the greenhouse due to drafts and "cold pockets".

Root Zone Temperatures

Root zone temperatures are primarily managed to remain in a narrow range of 18 to 21°C to ensure proper root functioning. Control of the root zone temperature is primarily a concern for winter zones, and is obtained through the use of bottom heat systems such as pipe and rail systems. Control is maintained by monitoring the temperature at the roots and maintaining the pipe at a temperature that ensures optimum root zone temperatures. The use of tempered irrigation water is also a strategy employed by some growers. Maintaining warm irrigation water (20°C is optimum) minimizes the shock to the root system associated with the delivery of cold irrigation water. In cases during the winter months, in the absence of a pipe and rail system, root zone temperatures can drop to 15°C or lower. The performance of most greenhouse vegetable crops is sub-optimal at this low root zone temperature. Using tempered irrigation water alone is not usually successful in raising and maintaining root zone temperatures to optimum levels. The reasons for this are two folds firstly, the volume of water required for irrigation over the course of the day during the winter months is too small to allow for the adequate sustained warming of the root zone, and secondly, the temperature of the irrigation water would have to be almost hot in order to effect any immediate change in root zone temperature. Root injury can begin to occur at temperatures in excess of 23°C in direct contact with the roots. The recommendation for irrigation water temperature is not to exceed 24-25°C.

During the hot summer months temperatures in the root zone can climb over 25°C if the plants are grown in sawdust bags or rockwool slabs, and if the bags are exposed to prolonged direct sunlight. Avoiding high root zone temperatures is accomplished primarily by ensuring an adequate crop canopy to shade the root system. Since larger volume of water is applied to the plants during the summer, ensuring that the irrigation water is relatively cool approximately 18°C, (if possible) will help in preventing excessive root zone temperatures. One important point to keep in mind with respect to irrigation water temperatures during the summer months is irrigation pipe exposed to the direct sun can cause the standing water in the pipe to reach very high temperatures as high as 35°C. Irrigation pipe is often black to prevent light penetration into the line which can result in the development of algae and the associated problems with clogged drippers. It is important to monitor irrigation water temperatures at the plant drip lines especially during the first part of the irrigation cycle to ensure that the temperatures are not too high. All exposed irrigation pipes should be shaded with white plastic or moved out of the direct sunlight if a problem is detected.

Relative Humidity

Plants exchange energy with the environment primarily through water evaporation by the process of transpiration. Transpiration is the only type of transfer process in the greenhouse that has both a physical and biological basis. This plant process is almost exclusively responsible for the sub-tropical climate inside the greenhouse. Light energy to the extent of 70% falling on a greenhouse crop goes towards transpiration making change of liquid water to water vapour and the most of the irrigation water applied to the crop is lost through transpiration. Relative humidity (RH) is a measure of the water vapour content of the air. The use of relative humidity to measure the amount of water in the air is based on the fact that the ability of the air to hold water vapour is dependent on the temperature of the air. Relative humidity is defined as the amount of water vapour in the air compared to the maximum amount of water vapour the air is able to hold at that temperature. The implication of this is that a given reading of relative humidity reflects different amounts of water vapour in the air at different temperatures. For example, air at a temperature of 24°C and a RH of 80% is actually holding more water vapour than the air at a temperature of 20°C and a RH of 80%.

The use of relative humidity for control of the water content of the greenhouse air mass has commonly been approached by maintaining the relative humidity below threshold values, one for the day and one for the night. This type of humidity control is directed at preserving low humidity, and although humidity levels high enough to favour disease organisms must be avoided, there are more optimal approaches to control the humidity levels in the greenhouse environment. The sole use of relative humidity as the basis of controlling greenhouse air water content does not allow for optimization of the growing environment, as it does not provide a firm basis for dealing with plant processes such as transpiration in a direct manner. The common purpose of humidity control is to sustain a minimal rate of transpiration. The transpiration rate of a given greenhouse crop is a function of three in-house variables namely temperature, humidity and light. Light is the one variable usually outside the control of most greenhouse growers. If the existing natural light levels are accepted, then crop transpiration is primarily determined by the temperature and humidity in the greenhouse. Achievement of the optimum "transpiration set point" depends on the management of temperature and humidity within a greenhouse. More specifically, at each level of natural light received into the greenhouse, a transpiration set point should allow for the determination of optimal temperature and humidity set points. The relationship between transpiration and humidity is awkward to describe, as it is largely related to the reaction of the stomata to the difference in vapour pressure between the leaves and the air. The most certain piece of knowledge about how stomata behave under increasing vapour pressure difference is it is dependent on the plant species in question. However, even with the current uncertainties with understanding the relationships and determining mechanisms involved, the main point to remember about environmental control of transpiration is that it is possible. The concept of vapour pressure difference or vapour pressure deficit (VPD) can be used to establish set points for temperature and relative humidity in combination to optimize transpiration under any given light level. VPD is one of the important environmental factors influencing the growth and development of greenhouse crops, and offers a more accurate characteristic for describing water saturation of the air than relative humidity because VPD is not temperature dependent. Vapour pressure can be thought of as the concentration, or level of saturation of water existing as a gas, in the air. As warm air can hold more water vapour than cool air, so the vapour pressures of water in warm air can reach higher values than in cool air. There is a natural movement from areas of high concentration to areas of low concentration. Just as heat naturally flows from warm areas to cool areas, so does water vapour move from areas of high vapour pressure, or high concentration to the areas of low vapour pressure or low concentration. The vapour pressure deficit is used to describe the difference in water vapour concentration between two areas. The size of the difference also indicates the natural "draw" or force driving the water vapour to move from the area of high concentration to low concentration. The rate of transpiration, or water vapour loss from a leaf into the air around the leaf, can be thought of and managed using the concept of vapour pressure deficit (VPD). Plants maintained under low VPD had lower transpiration rates while plants under high VPD can experience higher transpiration rates and greater water stress. A key point when considering the concept of VPD as it applies to controlling plant transpiration is the vapour pressure of water vapour is always higher inside the leaf than outside the leaf. Meaning the concentration of water vapour is always greater within the leaf than in the greenhouse environment, with the possible exception of having a very undesirable 100% relative humidity in the greenhouse environment. This means the natural tendency of movement of water vapour is from within the leaf into the greenhouse environment. The rate of movement of water from within the leaf into the greenhouse air, or transpiration, is governed largely by the difference in the vapour pressure of water in the greenhouse air and the vapour pressure within the leaf. The relative humidity of the air within the leaf can be considered to always be 100%, so by optimizing temperature and relative humidity of the greenhouse air, growers can establish and maintain a certain rate of water loss from the leaf, a certain transpiration rate. The ultimate goal is to establish and maintain the optimum transpiration rate for maximum yield. Crop yield is linked to the relative increase or decrease in transpiration, a simplified relationship relates increase in yield to increase in VPD.

Transpiration is a key plant process for cooling the plant, bringing nutrients in from the root system and for the allocation of resources within the plant. Transpiration rate can determine the maximum efficiency by which photosynthesis occurs, how efficiently nutrients are brought into the plant and combined with the products of photosynthesis, and how these resources for growth are distributed throughout the plant. Since the principles of VPD can be used to control the transpiration rate, there is a range of optimum VPDs corresponding to optimum transpiration rates for maximum sustained yield.

The measurement of VPD is done in terms of pressure, using units such as millibars (mb) or kilopascals (kPa) or units of concentration, grams per cubic meter (g/m3). The units of measurement can vary from sensor to sensor, or between the various systems used to control VPD. The optimum range of VPD is between 3 to 7 gram/m^3, and regardless of how VPD is measured, maintaining VPD in the optimum range can be obtained by meeting specific corresponding relative humidity and temperature targets.

The plants themselves exert tremendous influence on the greenhouse climate, transpiration not only serves to add moisture to the environment, but is also the mechanism by which plants cool themselves and add heat to the environment. Optimization of transpiration rates through management of air temperature and relative humidity can change over the course of the season. Early in the season, when plants are young and the outside temperatures are cold, both heat and humidity (from mist systems) can be applied to maintain temperature and humidity. As the season progresses and the crop matures, increasing light intensity increases the transpiration rate and the moisture content of the air. To maintain optimum rates of transpiration, ventilation is employed to reduce the relative humidity in the air. However, under typical summer conditions, ventilation is almost exclusively triggered by high temperature set points calling for cooling. Under these conditions, ventilation can occur continuously throughout the daylight period and results in very low relative humidity in the greenhouse. As the hot, moist air is vented, it is replaced by still warm, dry air. So, additional cooling through mist systems or fan and pad evaporative cooling is required to both reduce the amount of ventilation for cooling as well as to add moisture to the air.

Carbon Dioxide (CO_2)

Photosynthesis is the process which involves a chemical reaction between water and carbon dioxide (CO_2) in the presence of light to make food (sugars) for plants, and oxygen is released in the atmosphere as a by-product. Carbon dioxide currently comprises 0.04% (400 ppm) of the atmospheric volume. Plants take in CO_2 through small cellular pores called stomata in the leaves during the day.

During respiration (oxidation of stored sugars in plants producing energy and CO_2) plants take in oxygen (O_2) and give off CO_2, which complements photosynthesis when plants take in CO_2 and give off O_2. The CO_2 produced during respiration is always less than the amount of CO_2 taken in during photosynthesis process. So, plants are always in a CO_2 deficient condition, which limits their potential growth.

Although atmospheric and environmental conditions like light, water, nutrition, humidity and temperature may affect the rate of CO_2 utilization, the amount of CO_2 in the atmosphere has a greater influence. Variation in CO_2 concentration depends upon the time of day, season, number of CO_2 producing industries, composting, combustion and number of CO_2 absorbing sources like plants and water bodies. The ambient CO_2 (naturally occurring level of CO_2) concentration of 400 ppm can be achieved in a properly vented greenhouse. However, the concentration is much lower than ambient during the day and much higher at night in sealed greenhouses. The carbon dioxide level is higher at night because of plant respiration and microbial activities. The carbon dioxide level may drop to 150 to 200 ppm during the day in a sealed greenhouse, because CO_2 is utilized by plants for photosynthesis during day time. Exposure of plants to lower levels of CO_2 even for a short period can reduce rate of photosynthesis and plant growth. Generally, doubling ambient CO_2 level (700 to 800 ppm) can make a significant and visible difference in plant yield. Plants with a C3 photosynthetic pathway (geranium, petunia, pansy, aster lily and most dicot species) have a 3 carbon compound as the first product in their photosynthetic pathway, thus are called C3 plants and are more responsive to higher CO_2 concentration than plants having a C4 pathway (most of the grass species have a 4 carbon compound as the first product in their photosynthetic pathway, thus are called C4 plants). An increase in ambient CO_2 (800-1000 ppm) can increase yield of C3 plants up to 40 to 100% and C4 plants by 10 to 25% while keeping other inputs at an optimum level. Plants show a positive response up to 700 to need of 1800 ppm, but higher levels of CO may cause plant damage (**Figure 12.3**).

CO_2 Supplementation

The greenhouse industry has advanced with new technologies and automation. With the development of improved lighting systems, environmental controls and balanced nutrients, the amount of CO_2 is the only limiting factor for maximum growth of plants. Thus, keeping the other growing conditions ideal, supplemental CO_2 can provide improved plant growth. This is also called 'CO_2 enrichment' or 'CO_2 fertilization.'

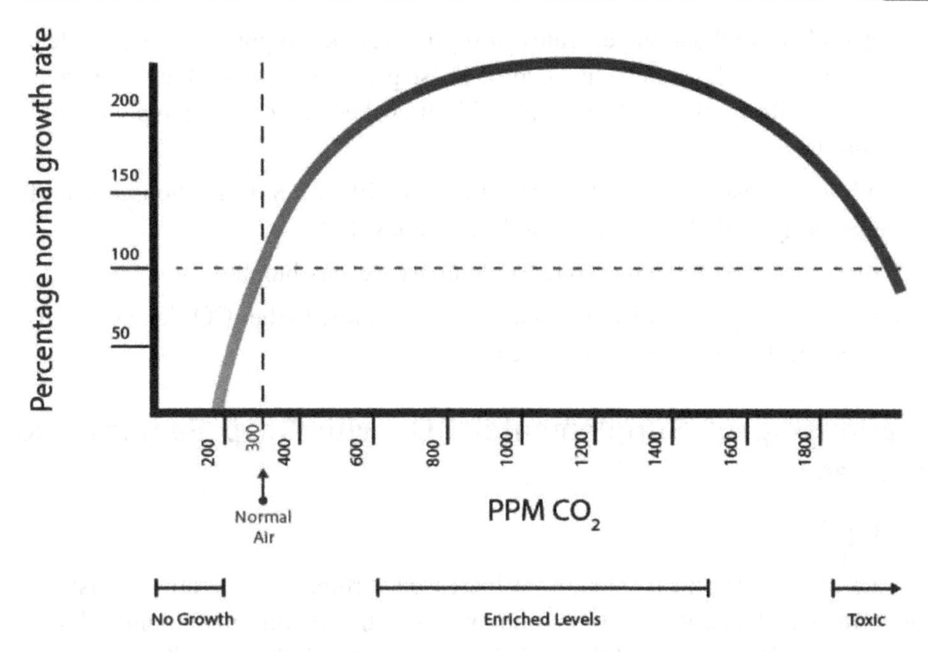

Figure 12.3: Relation between CO_2 concentration and rate of plant growth

Coustesy: Poudel and Dunn, 2017 & Roger, 2016

Advantages

1. Increase in photosynthesis results in increased growth rates and biomass production.
2. Plants have earlier maturity and more crops can be harvested annually. The decrease in time to maturity can help in saving heat and fertilization costs.
3. In flower production, supplemental CO_2 increases the number and size of flowers, which increase the sale value because of higher product quality.
4. Supplemental CO_2 provides additional heat (depending upon the method of supplementation) through burners, which will reduce heating cost in winter.
5. It helps to reduce transpiration and increases water use efficiency resulting in reduced water use during crop production.

Disadvantages

1. Higher production cost with a CO_2 generation system.
2. Plants may not show a positive response to supplemental CO_2 because of other limiting factors such as nutrients, water and light. All factors need to be at optimum levels.
3. Supplementation is more beneficial at younger stage of plants.

4. Incomplete combustion generates harmful gases like sulphur dioxide, ethylene, carbon monoxide and nitrous oxides. These gases are responsible for necrosis, flower malformation and senescence if left unchecked resulting in poor quality of produce.
5. Additional costs required for greenhouse modification. Greenhouses need to be properly sealed to maintain a desirable level of CO_2.
6. Excess CO_2 level can be toxic to plants as well as humans.
7. On warmer days, it is difficult to maintain desirable higher CO_2 levels because of venting to cool the greenhouses.

Relationship of Supplemental CO_2 with Crop Factors and Advantages

CO_2-light

The rate of photosynthesis cannot be increased further after certain intensity of light termed as the light saturation point, which is the maximum amount of light a plant can use. However, additional CO_2 increases the light intensity required to obtain the light saturation point, thus increasing the rate of photosynthesis. Mostly in the winter, photosynthesis is limited by low light intensity. An additional lighting system will enhance the efficiency of CO_2 and increase the rate of photosynthesis and plant growth. Thus, supplemental CO_2 integrated with supplemental lighting can decrease the number of days required for crop production.

CO_2-water

CO_2 supplementation affects the physiology of plants through stomatal regulation. Elevated CO_2 promotes the partial closure of stomatal cells and reduces stomatal conductance. Stomatal conductance refers to the rate of CO_2 entering and exiting with water vapor from the stomatal cell of a leaf. Because of reduced stomatal conductance, transpiration is minimized which results into an increased water use efficiency (WUE= water used in plant metabolism/water lost through transpiration). Lower stomatal conductance, reduced transpiration, increased photosynthesis and an increase in WUE helps plants to perform more efficiently in water stress conditions. Supplemental CO_2 reduces water demand and conserves water during water scarce periods of a specific location.

CO_2-temperature

Most of the biological processes in plants increase with increasing temperature, which is also applicable to the rate of photosynthesis. But, the optimum temperature for maximum photosynthesis depends on the availability of CO_2. The higher the amount of available CO_2, the higher the optimum temperature requirement of crops (**Figure 12.4**). In a greenhouse supplemented with CO_2, a dramatic increase in the growth of plants can be observed with increasing temperature. Supplemental CO_2 increases the optimum temperature requirement of a crop. This increases crop production even at higher temperature, which is not possible at the ambient CO_2.

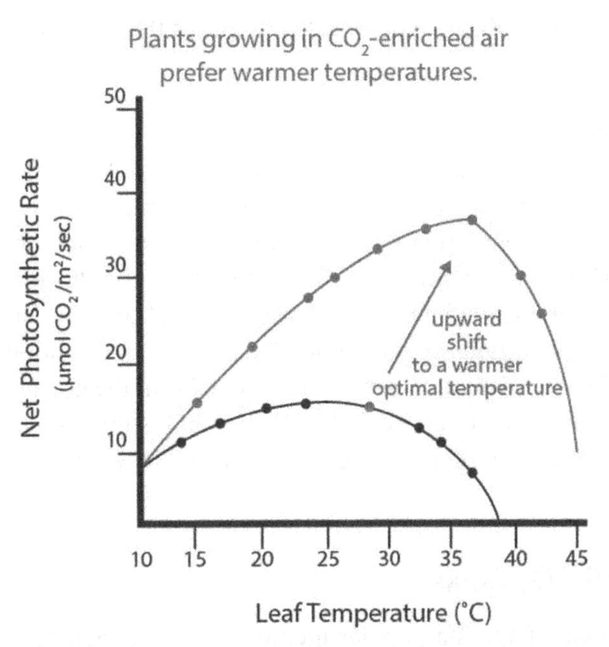

Figure 12.4: Relationship between leaf temperature and net photosynthetic rate at ambient and CO_2 elevated condition in *Populas grandidentata*

Coustesy: Poudel and Dunn, 2017 & Jurik *et al.*, 1984

CO_2 –nutrient

Plants show rapid growth as a result of CO_2 supplementation because of enhanced root and shoot growth. The enhanced root system allows greater uptake of nutrients from the soil, hence application of higher dose of fertilizers is recommended with the increasing level of CO_2. The normal fertilizer rate can be exhausted quickly and plants may show several nutrient deficiency symptoms. Although, strict recommendations of nutrients for different crops at different levels of CO_2 are not presently available, in general nutrient requirements increase with increasing levels of CO_2. On the other hand, some micro-nutrients are depleted quicker than macro-nutrients. Some studies have reported low levels of zinc and iron in crops produced at higher CO_2 levels. Further decrease in transpiration and conductance with CO_2 supplementation may affect calcium and boron uptake, which should be compensated through addition of nutrients.

Sources of Carbon Dioxide

Carbon dioxide should be supplemented in a pure form. A mixture of carbon monoxide, ozone, nitrogen oxides, ethylene and sulfur impurities in some CO_2 sources may damage the plant. Carbon monoxide should not exceed 50 ppm, otherwise CO_2 supplementation will be harmful rather than beneficial. There are different methods of CO_2 supplementation and the principle of CO_2 production is different depending on the method selected. Some of the methods are given below:

Natural CO_2

Since CO_2 is a free and heavy gas, it stays at a lower level in the greenhouse. Carbon dioxide produced by plants at night is depleted within a few hours after sunrise, thus proper ventilation integrated with horizontal airflow fans just above the plant can help in distributing available CO_2 at least to the ambient level. It is the cheapest method of maintaining an ambient level of CO_2. But in winter, the extreme climatic conditions do not favour this method and additional CO_2 sources are required. Another natural way of increasing CO_2 in the greenhouse is through human respiration. Humans also exhale CO_2 during respiration like plants. People working in the greenhouse for pruning, irrigation and other operations can increase CO_2.

Compressed CO_2 Tanks

Use of compressed CO_2 is a popular method of CO_2 enrichment. The CO_2 in a compressed liquid form vaporizes through use of CO_2 vaporizer and is distributed through a distribution system. Holes are added to PVC pipes and spread throughout the greenhouse for even distribution. However, CO_2 is released directly from a tank in small greenhouses. Generally, it is an expensive method in which liquefied CO_2 is brought by a large truck and put in storage tanks for larger operations but small tanks are also available for small growers. Along with the tank, a pressure regulator, flow meter, solenoid valve, CO_2 sensors and timers are required for the operation. Because of increased precision with compressed CO_2, most operators use advanced digital regulators. Other accessory costs are higher and makes the method quite expensive.

CO_2 Generator

Combustion of hydrocarbon fuels generally produces CO_2, water and heat. Greenhouse growers can use small CO_2 generators operated with propane or natural gas. Burning of approx 450 g fuel can produce 1350g of CO_2. 450g of

CO_2 is equivalent to 8.7 cubic feet of gas at standard temperature and pressure. At this rate, 28 g ethyl alcohol per day is required to maintain 1300 parts per million of CO_2 for a 200 square foot sized room. The amount of CO_2 produced depends on the type and purity of fuel. But, combustion without adequate oxygen may produce impurities which are harmful to plants. So, smaller areas should be opened for fresh air even in sealed greenhouse conditions. These generators are kept just above the plants and each unit covers about 4,800 square feet of area. The CO_2 burner capacity ranges from 20,000 to 60,000 Btu per hour and can produce 3.7 kg CO_2 per hour by burning natural gas. Instead of using small generators in multiple greenhouse bays, larger greenhouse operations use gas engines to produce flue gas (exhaust gas of engine), which passes through a series of filters to give pure CO_2. The main advantage of this system is that it produces both heat and electricity along with CO_2. Heat is stored in a tank in the form of hot water and will be used in heating the greenhouses at night. Such big generators are capable of minimizing heating and electricity cost.

Decomposition and Fermentation

Organic matter decomposed by microbial action produces CO_2. Organic waste can be decomposed in plastic containers and the CO_2 produced can be used by plants. However, this method may require more space and substrate to produce adequate CO_2. Although it is an inexpensive method, but it is hard to control the concentration of CO_2 and gives off bad odours. To eliminate these disadvantages, many commercial products have been introduced in the market. The CO_2 boost bucket, Pro CO_2 and Exhale mushroom bag are some commercial products which claim to produce the desired level of CO_2 without odours. They could be beneficial for small scale growers and indoor gardens. Carbon dioxide is also a by-product of fermentation. Some growers use sugar solution and yeast to supplement CO_2. 450g sugar produces approximately 225 g ethanol and 225 g of CO_2. A suitable size plastic container, sugar, yeast and a sealant (to seal the container tightly) are necessary to start the production of CO_2. This method provides CO_2 faster than decomposition but has the disadvantages of foul odours, difficulty in maintaining desired concentrations and occupying a larger space. The major advantage of this method is the ethanol production. Ethanol is an organic fuel and can produce more CO_2 when burned.

Dry Ice

Dry ice is one of the cheapest methods adopted by growers in smaller greenhouses. In advanced greenhouses, special cylinders with a gas flow meter are used to control CO_2 regulation through sublimation of dry ice. Dry ice is a solid state of

CO_2 obtained by keeping CO_2 at an extremely low temperature (-78ºC). Slow release of dry ice may help in cooling greenhouses by a few degrees in the summer. In general, about 450g dry ice is sufficient to maintain 1300 ppm CO_2 in a 100 square foot area throughout the day. In a normal greenhouse, dry ice is sliced into small pieces and replaced every two hours to maintain a desired level of CO_2 or kept inside an insulator with small holes through which CO_2 escapes. It is cheap, readily available and can last for a whole day. The major disadvantages are low shelf-life and difficulty in storing at normal conditions. Rapid sublimation of dry ice may lead to increase level of CO_2 higher than 2000 ppm, which could limit growth as well could be toxic to plants.

Chemical Method

The chemical reaction of baking soda with acid (mostly acetic acid) can produce CO_2, but a large quantity of materials is required to produce adequate CO_2. Reaction of about 900g baking soda with 10 to 12 litres of 5% acetic acid produces only 450g CO_2. Thus, this is considered an expensive method of CO_2 production.

Things to ponder during CO_2 Supplementation

1. Never allow CO_2 to exceed the plant requirements. Keep an alert system when CO_2 level reaches 2000 ppm, because a high level of CO_2 (5000 ppm and above) can kill people.
2. Always monitor the CO_2 levels through sensors and adjust to required level.
3. Use a pure form of CO_2 and provide enough oxygen for combustion for the elimination of toxic gases.
4. Always keep the CO_2 source above the plant (except in the flue gas system) and evenly distribute the air inside the greenhouse.
5. Choose the method of supplementation that suits individual's operations. Develop a strategy based on a cost/benefit analysis. Maintain ideal growing condition like proper lighting, moisture, temperature, nutrition and humidity to make CO_2 supplementation effective.

13

PRECONDITIONS FOR CROPS AND VARIETIES SELECTION UNDER PROTECTED CULTURE

The choice of crops and cultivars is one of the deciding factors governing the outcome from greenhouse production system and various points like agro-ecological constraints, technological developments and socio-economic opportunities must be taken into consideration while taking protected cultivation as commercial venture. The choice of crops and varieties for protected cultivation is considered as basic variable that affect the success and economic gain from production programme significantly. Good Agricultural Practices for protected cultivation also recommend the best choice of genotypes as priority factor to harness the full potential of greenhouse technology. Greenhouse production is claimed to be an intensive production system, which allows a very narrow window in between crops cycles. So, greenhouse growers must keep in mind 4 important points for getting success from the system and these are 1) which crop should be grown depending upon the demand 2) when is the appropriate time to produce that crop 3) What are the different horticultural practices and requirements of the crop 4) Grower must have an idea of disposing the produce for higher returns (market analysis).

Considerations for Crops Selection under Protected Culture

- Select an economically potential crop of the region, which a grower must have complete knowledge of horticultural operations like training and pruning, fertigation *etc*.
- Select a crop appropriately suited to the type of greenhouse structures, if an option of multiple structures is available with the grower.

Identification of suitable market is always a challenge for growers practicing intensive production programme like protected cultivation for the year round supply. Higher yield per unit greenhouse area can be obtained following all good horticultural practices, however sometimes production doesn't meet the market demand which is directly reflected in the economic gain from the produce.

The size of farm also plays very important role in deciding the selection of crop and variety, as in general, medium and large farms may sell their produce at distantly placed domestic as well as in international markets. The growers with small sized greenhouses have very limited choice and can think of local markets for any economic return. Therefore, it is very important for growers dealing local markets, national or international markets to select targeted crops as per the demand in the respective destinations.

Vegetable crops, cut flowers, ornamentals and even few fruit crops may be important for cultivation under protected environments, but geographical adaptation of a particular crop may not necessarily guarantee the success of that crop under protected environment because of climatic factors as well as technological reasons.

Choice of the Crop

As protected cultivation stipulates the concept of off-season production of crops for higher economic gain, so selection of any of the crops from the list of vegetable crops like tomato, bell pepper, brinjal, cucumber, musk melon, summer squash, water melon, leafy vegetables *etc*., flower crops like Rose, Gerbera, Orchids, Anthurium, Heliconia, pot ornamental plants and fruit crops like strawberry, papaya *etc*. must target the supply of crops to the market during lean period when there is no or negligible competition from outdoor cultivation. The success in protected cultivation of any of the crops is dependent on the following factors other than off-season targeted production:

- Extensive demand and use of produce/product.
- Good adaptability under greenhouse environment.
- Extended cropping span for continuous use of protected structures during the year.

The choice of a specific crop has become interestingly very important for economic sustainability of greenhouse production system and can play a vital role in channelizing the production levels in new greenhouses being fabricated in new areas.

The choice of a crop must include species and varieties with higher yield potential fulfilling market demand to achieve maximum remuneration. The compatibility of crop with prevailing micro-climate of protected structure available with a grower and soil characteristics, and soil-borne diseases are also given considerable weightage during the selection of a particular crop with more specific consideration as listed below:

- Accessibility to the markets
- Plant architecture
- Crop specific requirements
- Skilled-unskilled labour requirements
- Feasibility for climate control (both active and passive)

Market prices hold strong relationships with growers' returns, as most of the component of production costs remains variable, for instance different wages are given to skilled labour, unskilled labour. Similarly, variable components of cost like fertilizers, pesticides, transportation etc. do vary over a period of time. So, when a grower is targeting local market, transportation cost can be reduced significantly by arranging road side markets or making small arrangements with local grocery stores. However, these systems are not very common for greenhouse produce and greenhouse growers mostly reckon centralized market distribution or arrangements with supermarkets and supermarket chains. The cold storage facilities at farm level if possible or locally could be very critical in preserving the quality of perishable produce prior to transportation.

Choice of Suitable Variety

Selection of type of varieties in different crops must be given due consideration while going for their cultivation under protected environment. A particular variety of a specific crop might be performing excellently under outdoor conditions in a region, but the same variety when cultivated inside the greenhouse environment may not produce higher yields as per the anticipation. For instance, monoecious type of cucumber cultivars perform very well under open conditions, however they are very poor yielders under protected environment because of their requirement for pollination. Even if a grower ensures artificial pollination inside protected structures through natural pollinators like honey bees, but yield potential of these types of varieties doesn't meet with that of specific greenhouse varieties (parthenocarpic varieties of cucumber). Specific requirements of different cultivar may also exist depending on the level of technology used in protected cultivation, for example indeterminate type of varieties should be selected for greenhouse cultivation owing to their long crop cycle.

Traders consider long shelf-life of a produce as one of the important factors along with uniqueness and appreciation of product by the consumers for overall economic gain. While the consumers look the products from different perspectives and these should be versatile, must have good taste and free from any contaminants like pesticide residues.

The advances in breeding have resulted in developing specific varieties substantially with improved disease resistance, adaptability to suboptimal temperature and light, and other specific traits such as parthenocarpy and suitability for grafting. Now special attention is paid to define qualitative traits defining the nutritional profile of fresh fruits and vegetables additionally to other standard quality parameters like size, colour, TSS, dry matter, shelf-life *etc.*

Screening of newly identified varieties must be carried out locally in close association with local administrations and research institutions to identify and endorse potential varieties for the region. Similarly, transfer of technologies on different aspects of protected cultivation of crops must be ensured for harnessing full potential of greenhouse technology.

Ultimately, the fundamental requirement for greenhouse crops should meet year round production, adaptation to soil as well as soilless culture, high yield potential, low labour requirements, high thermal requirements (making open-air cultivation impossible) and high quality. While making introduction of any new crop in the region for cultivation under greenhouse environment, it should fulfill the requirements of their adaptation to new agro-climatic and social conditions, must fit well to consumer requirements and should be worth marketable and profitable.

GAP Recommendations for Choice of Crops and Varieties under Protected Culture:

1. The physiological requirements of a crop and its varieties in relation to light, temperature, relative humidity and soil parameters of given environment must be taken into consideration.
2. The choice of a crop and its specific variety for a given environment must also be validated in context to economics.
3. Transfer of technologies (ToT) by arranging short training, demonstrations on/off the farm, skill development trainings must be ensured for proper and better adoption of technology.
4. The growers must also be made aware of new market trends in relation to current economic convenience.
5. The advantages and disadvantages of diversification compared with specialization on a case to case basis should be considered.

6. Crop diversification and introduction of new crops or varieties should be adopted in a coordinated manner so as to capitalize on available labour, technical specialization, environmental parameters and socio-economic constraints in a specific area.
7. The cost-benefit analysis of production system shouldn't be restricted to agronomic evaluation, but socio-economic parameters and time should also be reviewed critically.

14

BASIC CONSIDERATIONS FOR SITE SELECTION, ORIENTATION AND DESIGNING OF GREENHOUSE STRUCTURES

Greenhouse designing should be based upon sound scientific principles so as to create controlled or semi-controlled or partially controlled environment in favour of crop growth and development. There are some important aspects which need to be given adequate attention while starting protected cultivation.

Site Selection

While selecting a site for greenhouse, the following points should be considered for optimum growth and development of plants:

- The site should be free from shadow.
- The site should be at a higher level than the surrounding land with adequate drainage facility.
- Availability of good quality irrigation water and electricity to run the fan & pad cooling system.
- pH of the irrigation water should be in the range of 5.5 to 7.0 and EC between 0.1 to 0.3 $mScm^{-1}$.
- pH of the soil should be in the range of 5.5 to 6.5 and EC between 0.5 to 0.7 $mScm^{-1}$.
- Proximity to motorable road to take advantage of market for inputs supply and sale of produce.
- Soil needs to be sterilized after every 3 to 4 years to avoid built up of soil-borne pathogens particularly nematodes or one may resort to soil-less cultivation.

Orientation

An accurate orientation of protected structure can provide congenial environmental conditions for growth and development of crops grown inside the greenhouse. Following points should be examined while deciding the orientation of a greenhouse depending upon light intensity, and direction and velocity of wind. The pictorial depiction of greenhouse orientation is shown in **Fig. 14.1**.

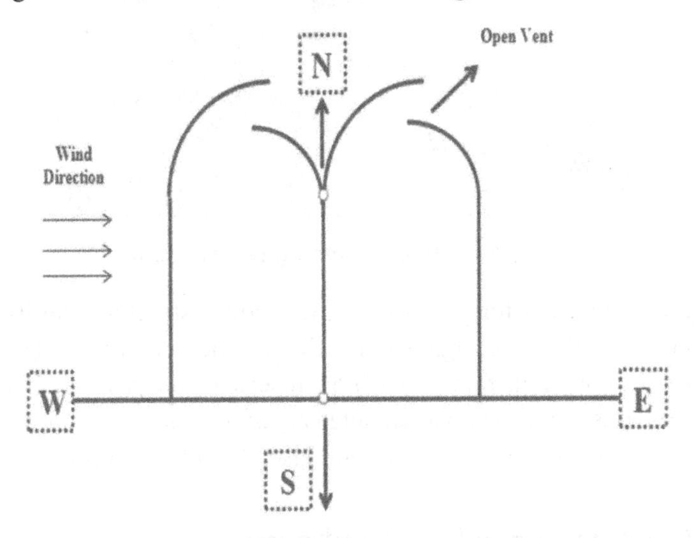

Fig. 14.1: Greenhouse Orientation

- Orientation of a single span greenhouse should be directed towards East-West (length to width), while it should be in North-South direction for multispan type greenhouses (gutter direction).
- Slope along the gutter should not be more than 2%, while it should not be more than 1.25% along the gable side.
- Ventilators should open on the leeward side in naturally ventilated greenhouse (**Fig. 14.2**).
- Single span greenhouse should have long axis perpendicular to the direction of wind to protect it from wind damage.
- Wind breaks should be placed at least 30 m away on North-West side of the greenhouse.
- There is less availability of solar radiation and the sky remains cloudy during rainy season in tropical regions, hence so effective ventilation is very important to manage temperature and humidity by keeping proper orientation of polyhouse.

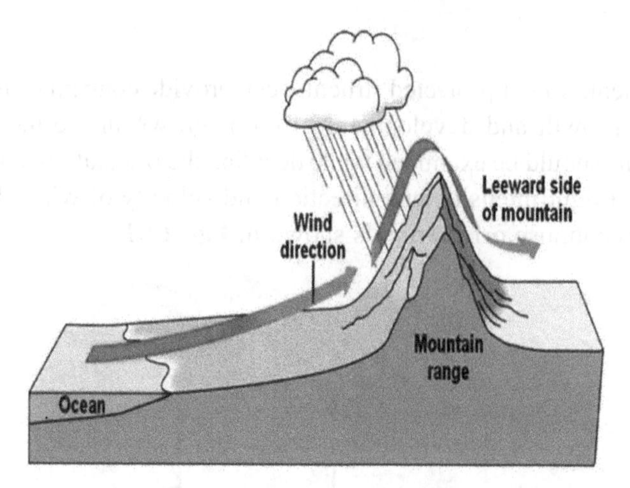

Fig. 14.2: Understanding Leeward side

The most important function of the greenhouse structure and its covering is the protection of the crop against hostile weather conditions (low and high temperatures, snow, hail, rain and wind), diseases and pests. It is important to develop greenhouses with a maximum intensity of natural light inside. The structural parts that can cast shadows in the greenhouse should be minimized.

Design Considerations for Greenhouse

Hanan (1998), and Van Heurn and Van der Post (2004) have recommended some factors that determine the particular choice for protected cultivation system. A combination and extension of their lists of factors are:

1. Market size and regional physical and social infrastructure which determines the opportunity to sell the products as well as the costs associated with transportation.
2. Local climate which determines crop production and thus the need for climate conditioning and associated costs for equipments and energy. It also determines the greenhouse construction dependent of wind forces, snow and hail *etc.*
3. Availability, type and costs of fuels and electric power to be used for operating and climate conditioning of the greenhouse.
4. Availability and quality of water.
5. Soil quality in terms of drainage, the level of water table, risk of flooding and topography.
6. Availability and cost of land, present and future urbanization of the area, the presence of industries and zoning restrictions.

7. Availability of capital.
8. The availability and cost of labour as well as the level of education/ skills.
9. The availability of materials, equipment and service level that determines the structures and instrumentation of the protected cultivation systems.
10. Legislation in terms of food safety, chemical residues, the use and emission of chemicals to soil, water and air.

The design procedures roughly contain the following steps (Van den Kroonenberg and Siers, 1999):

1. Definition of design objective.
2. In a brief of requirements, the specifications and design objectives are stipulated. One may think of performance in terms of energy use, emission levels, labour requirements *etc.*
3. A system analysis will reveal the functions needed such as cooling, heating, nutrient and water supply, internal transport.
4. Deviation of alternative working principles for each function which yields so called morphological diagrams. For example, in case of cooling one may consider natural ventilation, forced ventilation, fog systems and fan and pad cooling as design alternatives.
5. Concept development stage, during this stage the different functions or more specifically working principles in the morphological diagram are combined into a conceptual design that should be at least satisfy the functional requirements stated in the design specifications. Several different concepts can be designed at this stage.
6. Design evaluation and bottle neck assessment. During this stage, the various conceptual designs are evaluated in the view of the design requirements stated above. The design evaluation is based on expert assessment but also on quantitative simulation using mathematical models. Bottle necks and contradictions in the design can be identified and hence one or two conceptual designs are chosen.
7. For the conceptual design (s) chosen, each working principle has to be worked out in more detail. Solutions for a bottle neck function and design contradictions have to be found.
8. The design prototype is built and tested in view of the design requirements.

The structural engineering aspect of protected structure has is to be dealt with utmost care, so that it must have very good strength to carry the following loads.

a) **Dead load**: Weight of all permanent construction, cladding, heating and cooling system, water pipes and all service equipments fixed to the frame.

b) **Live load**: Weights laid over by use of components like hanging baskets, shelves and persons working on top of the structure. The greenhouse has to be designed to carry live load maximum to 15 kgm^{-2}. Each member of roof should be capable of supporting 45kg concentrated load when applied at its centre.

c) **Wind load**: The structure should be able to withstand wind speed of 150 km hr^{-1} and at least 50 kg m^{-2} wind pressure.

d) **Snow load**: It is to be taken as per the average snowfall of the location. The greenhouse should be able to take dead load plus live load or dead load plus wind load plus half the live load.

The greenhouses are to be fabricated of Galvanized Iron (GI) Pipes. The foundation can be 60 cm x 60 cm x 60 cm or 30 cm diameter and 1 m depth in plain cement concrete (PCC) of 1:4:8 (one part of cement, four parts of sand and eight parts of coarse aggregates) ratio. The vertical poles should also be covered to the height of 60 cm by PCC with a thickness of 5cm, which will avoid the rusting of the poles.

Size

The size of the greenhouse needs to be determined based on the availability of land and the cost, which may vary depending upon the types of structure specifically fabricated for the region. Depending upon the market access and experience of greenhouse cultivation, it is suggested to start with a naturally ventilated greenhouse of minimum size as it would require less initial capital investment along with operational expenditure. However, experienced farmers/entrepreneurs may decide to go for larger greenhouse depending on their scale of operation and project costs.

Height

Height is one of the most important aspects of greenhouse design. The height of the structure directly impacts natural ventilation, stability of the internal environment and crop management. The ideal central height and side/gutter height of naturally ventilated greenhouse as well as for greenhouse with fan & pad system is give below:

Component	Type of structure		
	NVPH (up to 250 m^2)	NVPH (> 250 m^2)	Fan & Pad system
Central height	3.5-4.5 m	5.5-6.5 m	<5.5 m
Side/ gutter height	2.5-3.0 m	4.5-5.0 m	As per NVPH size

Both types of greenhouse can be made in single or multispan structures. A multispan greenhouse can be constructed for an area more than 200 m^2 and is economical in terms of construction material and required control/ monitoring equipment.

Cladding Materials

Cladding (Covering/glazing) materials are the most important constituent of protected structures as they have direct influence on micro-climate inside the greenhouse. Transmission for global radiation (Ultra Violet-UV and Photosynthetically Active Radiation-PAR), heat radiation (Near Infra Red-NIR and Far Infra Red-FIR), insulating effect, sensitivity to ageing (mainly UV degradation), permeability for humidity (water), mechanical strength (tensile and impact), fire behaviour, investment costs, available dimensions *etc.* are the important properties, which must be addressed adequately while evaluating cladding materials for a greenhouse. So, there are different types of cladding materials, which can be used in greenhouses depending upon the suitability to the region and project cost.

Glass as Cladding Material

Glass is preferred by many growers as covering material for greenhouses considering greenhouse technology a long term investment **(See details in chapter 10)**.

Flexible Plastic Films as Cladding Material

Flexible plastic films namely polyethylene, ethylene tetrafluoroethylene (Tefzel), polyvinyl chloride (PVC) and polyester are used as covering material in polyhouses **(See details in chapter 10)**.

As per the Indian Standard (IS 15827: 2009), the plastic films should have following important characteristics:

Important characteristics of covering materials:

Sr. No.	Type of plastic films	Characteristics	Uses
1.	Normal film	a) Transparency more than 80% b) Low Greenhouse Effect	Forcing and Semi-Forcing crops
2.	Thermic clear film	a) Good Transparency b) High infrared (IR) Effectiveness	As normal film, when greater infrared (IR) effectiveness is desired.
3.	Thermic diffusion	a) Diffusion Light b) High infrared (IR) Effectiveness	As normal film, when greater infrared (IR) effectiveness and Diffusing light is desirable.

Rigid Plastic as Cladding Material

Fiberglass-reinforced plastic (FRP) rigid panel, polycarbonates, and acrylics are used as rigid plastic coverings. Generally, rigid plastics have very good light transmission, but it usually decreases over a period of time because of aging and yellowing of coverings due to the impact of UV radiations **(See details in chapter 10)**.

Agro Shade Nets as Cladding Material

A wide range of agro shade nets are available in different shading percentages or shade factor *i.e.* 15%, 35%, 40%, 50% 75% and 90%, which are defined on the basis of percentage of shade delivered to plants growing under them **(See details in chapter 10)**.

Comparison among different cladding materials:

Type	Durability (year)	Transmission		Maintenance
		Light (%)	Heat (%)	
Polyethylene	01	90	70	Very high
Polyethylene UV resistant	02	90	70	High
Fibre glass	07	90	05	Low
Tedlar coated Fibre Glass	15	90	05	Low
Double strength glass	50	90	05	Low
Poly carbonate	50	90	05	Very Low

Criteria for Selecting Greenhouse Structure

Galvanized iron (GI) pipe based greenhouse must conform to ISI Trade mark with minimum 2 mm wall thickness

The structure should have the following features:

- Must have stub type-anchoring foundation, galvanized and nut bolted structure having options for future expansion.
- Should be strong enough to withstand various types of load such as live load, dead load, crop load, wind and snow load.
- G.I. gutter should be in a single piece having width of 500 mm and thickness of 1 mm.
- Must have aerodynamic shape along all peripheries to resist high wind velocity.
- The structure must be designed in such a way that it can resist wind velocity up to 150 Km hr^{-1}.
- There must be provision for fixing of UV stabilized plastic film with Aluminum/ G.I. profiles and zigzag spring lock.
- Should be strong enough to support the load of internal service systems and plant foliage.
- UV stabilized plastic film of 200 micron thickness conforming to Indian Standard IS 15827: 2009 should be used.
- Should be easy to service, cover and recover with cladding material.
- Should be easy to operate and must be cost-effective.

15

CLIMATE CONTROL SYSTEMS FOR COOLING AND HEATING IN GREENHOUSES

Cooling is considered as the basic necessity for greenhouse crop production in tropical and sub-tropical regions to overcome the problems of high temperatures during summer months. Development of suitable cooling system that provides congenial micro-climate for crop growth is a difficult task as the design is closely related to the local environmental conditions. Moreover, selection of appropriate technology for cooling depends on the choice of the crops to be grown, maintenance, ease of operation and economic viability. Evaluation of micro-climate in different designs of the greenhouse and establishing physical and physiological relationships of crops is necessary for the greenhouse designers to improve cooling system suitable for crop growth. Different techniques can be employed for cooling during summer months and these are:

Cooling Systems for summer

1. **Ventilation:** Natural ventilation is the direct result of pressure differences created and maintained by wind or temperature gradients. It requires less energy and equipment and is the cheapest method of cooling a greenhouse, hence also known as passive cooling system. It depends heavily on evapo-transpiration cooling provided by the crop. Natural ventilation may be used to take advantage in moderately warm climates or in hot, arid climates depending upon the availability and dependability of the wind. Crops capable of high rates of transpiration should be chosen to maximize evapo-transpiration cooling.

   Some greenhouses can be ventilated using side and ridge vents, which run through the full length of greenhouse and can be opened as needed to provide the desired temperature. This method uses thermal gradients creating circulation due to warm air rising. Naturally ventilated greenhouses are typically provided with vent openings on both sides of the ridge and on both side walls. Vent operation should be such that the leeward vents are opened to produce a vacuum at the top of the ridge. Opening of the windward vents produces a positive pressure in the house which is typically less efficient than vacuum operation. Pressure gradients are typically small so that large vent openings are necessary to provide adequate ventilation. The combined sidewall vent area should equal to the combined ridge vent area, and each should be at least 15 to 20% of the floor area. Ridge vents should be top-hinged and should run continuously through the full length of greenhouse. The vents should form a 60°angle with the roof when fully opened.

Open roof designs may eliminate the need for side or end wall roof vents when more than 50% of the roof area is open. Any natural ventilation system should have the means to open partially or fully in response to inside temperature with the recommendation (if feasible) with automatic control of such systems for better results. Automatic vent systems should be equipped with rain and wind sensors to permit closing during periods when crop or ventilator damage might occur.

Wind and buoyancy effect are the main controlling agents for getting favourable micro-climate and air exchange rate in tropical and sub-tropical environments. Buoyancy effect cannot be neglected completely, though its effect on natural ventilation is not prominent when the external wind speed exceeds 7 to 9 km hr^{-1}. Relatively larger ventilation openings provided at the roof and ridge can realize acceptable natural ventilation cooling in tropical and sub-tropical regions.

2. **Evaporative Cooling:** Evaporative cooling is a process that reduces air temperature by evaporation of water into water vapours. As the water evaporates, energy is lost from the air causing its temperature to drop. Two temperatures are important when dealing with evaporative cooling systems; dry bulb temperature and wet bulb temperature. Dry bulb temperature refers to air temperature and is measured by a regular thermometer exposed to the air stream. Wet bulb temperature is the lowest temperature that can be reached by the evaporation of water only. Unlike dry bulb temperature, wet bulb temperature is an indication of the amount of moisture in the air.

   Wet bulb temperatures can be determined by checking with local weather station or by investing in an aspirated psychrometer, a sling psychrometer or an electronic humidity meter. Wet bulb psychrometers consist of two thermometers exposed to the same air stream. The end of one thermometer is covered by a wetted wick. As the water in the wick evaporates, the temperature of the thermometer decreases to the wet bulb temperature. The other thermometer is exposed directly to the airstream and measures the dry bulb temperature. A sling psychrometer is mounted on a swiveled handle and whirled rapidly, while an aspirated psychrometer uses a small fan to provide air movement.

The best time to measure wet bulb temperature to calculate the potential cooling performance of the evaporative cooling system is in the afternoon. This is when dry bulb temperature is at its peak because solar radiation and outside temperatures are at their peak. This is also when the difference between dry bulb temperature and wet bulb temperature is greatest and there is maximum potential for cooling.

The most common way of accomplishing evaporative cooling in a greenhouse is through fan and pad system. High pressure fog systems are also used to cool the greenhouses. Because of involvement of energy for operation of these both the systems, they are termed as active cooling systems

Fan and pad evaporative cooling systems

Fan and pad system consists of exhaust fans at one end of the greenhouse and a pump circulating water through and over a porous pad (**Figure 15.1**) installed at the opposite end of the greenhouse. If all vents and doors are closed when the fans operate, air is pulled through the wetted pads and water evaporates. Removing energy from the air lowers the temperature of the air being introduced into the greenhouse.

The air will be at its lowest temperature immediately after passing through the pads. As the air moves across the greenhouse to the fans, the air picks up heat from solar radiation, plants, and soil and the temperature of the air starts increasing gradually. The resulting temperature increase as air moves down the greenhouse produces a temperature gradient across the length of the greenhouse, with the pad side being coolest and the fan side warmest.

Fog or Mist System

This system is based on spraying water as small droplets (diameter: 2-60 mm) with high pressure nozzles. Cooling is achieved by evaporation of droplets. Fogging can also be used to increase the relative humidity apart from cooling the greenhouse (**Figure 15.2**).

Fog is generally produced using a high pressure pump to atomize water by forcing it through fixed nozzles attached to the tips of rotating fan blades or similar arrangement. Atomization may also be accomplished by a rotating disk or acoustic oscillator. The atomized water droplets should be 0.5 to 50 ìm to ensure proper cooling. Fog may be distributed through: 1) fixed nozzles located appropriately throughout the house, 2) perforated poly tubes 3) horizontal air circulation systems 4) oscillating fan nozzle designs. Good quality water is needed for fogging systems. The water should be free of precipitates and salts, and filtered wherever necessity is felt. In naturally ventilated greenhouses, nozzles should be uniformly spaced throughout the house. Pipes and nozzles should be located over aisles to prevent water dripping directly onto plants. In fan cooled houses, most of the fog should be concentrated near the inlet with a small amount distributed evenly over the rest of the house.

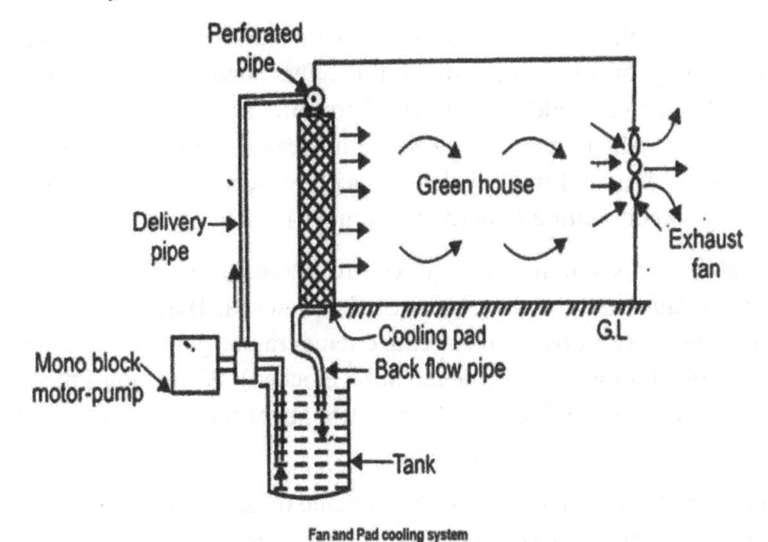

Fig. 15.1: Fan and Pad cooling system

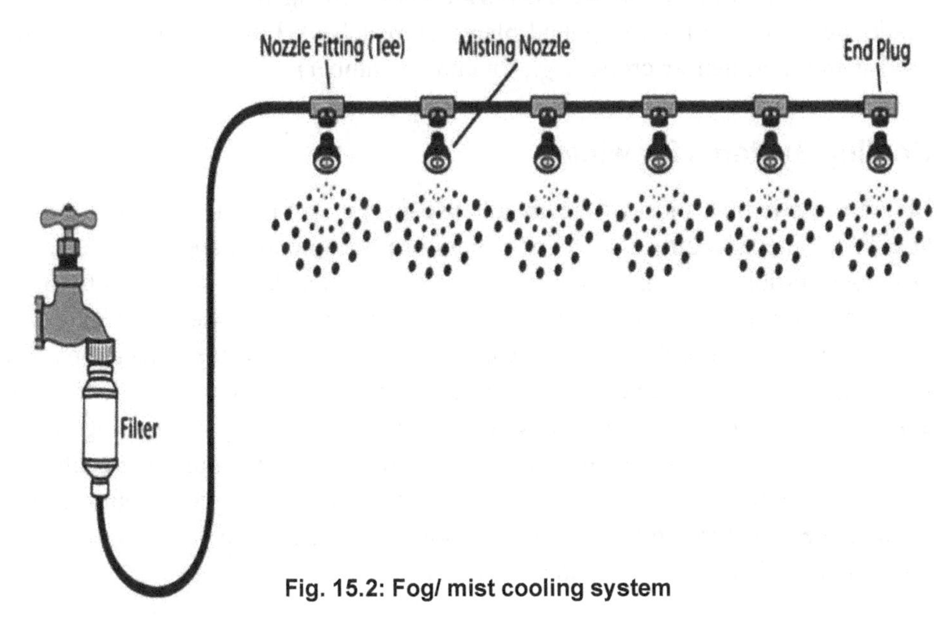

Fig. 15.2: Fog/ mist cooling system

Roof Evaporative Cooling

Roof evaporative cooling is achieved by sprinkling water onto the surface of greenhouse roof so as to form a thin layer of the free water surface causing increase in the evaporation rate and to fall the wet bulb temperature of the closely surrounded air.

Shading

Shading is a viable option for crop production during peak summer as it can control the entry of unwanted radiation or light. Shading paints/compounds can also be applied on the cladding material of greenhouse with some degree of success but paints get washed away during the rainy season. Shade net application with different mesh sizes and their evaluation with respect to local climate and region is necessary to get cooling benefits in summer months.

The ability of shade materials to control temperature is limited by the solar and thermal radiation characteristics of the material. Black and green external shade nets have been observed to reduce temperature gains by less than 50% of the shade rating (amount of visible radiation blocked). White cloths tend to provide greater temperature reduction. Very little information is available on the performance of Aluminized materials.

Where shade nets cannot be easily retracted during periods of cloudy weather, shade ratings of these nets used should be chosen carefully to avoid limiting growth. Higher values can be used with shade loving plants or crops with a shallow canopy depth (*e.g.*, potted plants on benches). Lower values should be used with taller, denser crops (*e.g.*, tomato, cucumber).

Cooling Systems for winter

In temperate region, most of the greenhouse designs are meant to conserve solar heat inside the structures. However, excess heat sometimes trapped inside could pose lot of problem causing injury to plants growing inside the structures because of the structural designs intended to serve the purpose of off-season production in such regions. Natural ventilation cannot be a good option for reducing excess heat from greenhouse structures in temperate regions as it may sometime cause extreme cooling if temperature outside is very low. So, the systems using active source of energy for cooling are suitable for brining the temperature of greenhouse structures down to a congenial level. The most commonly methods deployed for cooling the structures during winter seasons are convection tube cooling and horizontal air flow (HAF) fan cooling systems.

Convection Tube Cooling

The convection tube cooling system is comprised of a louvered air inlet, a polyethylene convection tube with air distribution holes. The system is also equipped with a pressurizing fan to direct air into the tube under pressure and an exhaust fan to create vacuum. The exhaust fan starts its function by creating vacuum inside the greenhouse structure when the air temperature inside exceeds the set point. The louvered air inlet gets opened due the vacuum created by exhaust fan and cool air start entering inside clear polyethylene convection tube with the help of pressurizing fan. Thus, cool air entering into the convection tube is dispersed by distribution holes present in it (**Fig. 15.3**).

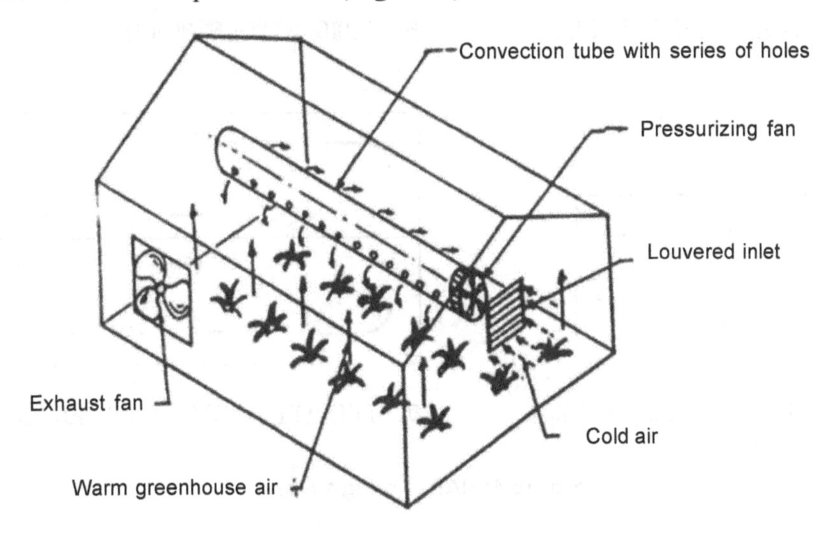

Fig. 15.3: Convection tube cooling system

Horizontal Air Flow (HAF) Cooling

HAF cooling system is proposed to an alternative to convection tube cooling system for even distribution of air inside greenhouse structures. The system uses small horizontal fans for moving the air mass and the greenhouse is considered as a large box containing air and the fans arrangements moves the air in a circular pattern systematically with a pressure @ 0.6 to 0.9 $m^3/min/m^2$ of the greenhouse floor area. It has been observed that the fans with a blade diameter of 41 cm and fractional horse power of 31 to 62 W (1/30 to 1/15 hp) could serve the purpose for effective air movement. The fans are generally positioned at 0.6 to 0.9 m above the plant canopy with 15 m distance intervals. The arrangement of fans is planned in such a way that the air flow is directed by one row of the fans along the length of the greenhouse down one side to the opposite end and then back along the other side by another row of fans. The number of fans to be used in this system depends largely on the size of greenhouse structures (**Fig. 15.4**).

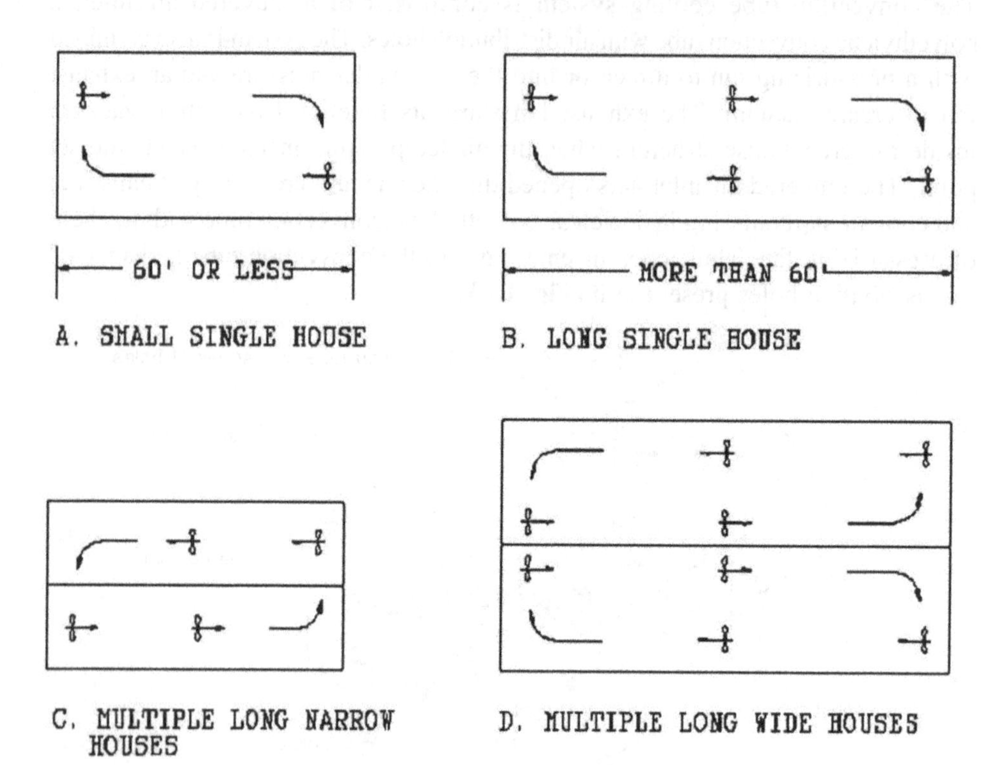

Fig. 15.4: HAF cooling system

Computers and Micro-controls

Thermostats and humidistats are reliable and inexpensive but are limited to simply turning equipments on or off in response to a change in temperature or relative humidity. Simple on-off controls cannot regulate environmental conditions exactly because they cannot sense how far the temperature or relative humidity is from the set point, or how rapidly the temperature or humidity is changing. Computers and microcontrollers can use software or hardwired circuits that incorporate logic to make decisions about the exact amount of heat or air flow required to produce desired environmental conditions inside the greenhouse.

Computers and microcontrollers are rapidly decreasing in cost, while at the same time increasing in reliability and sophistication. They are now important tools that growers can use to improve crop quality and increase profits. Computer based control systems can be linked to phone systems or to the Internet to allow growers to monitor greenhouse conditions closely from any location. Computer systems can also keep continuous records of greenhouse conditions and can be used to send messages or alarms to greenhouse growers when environmental conditions are out of range or when equipment fails. The increased control provided by these devices results in greenhouse conditions that provide a better environment for crop growth.

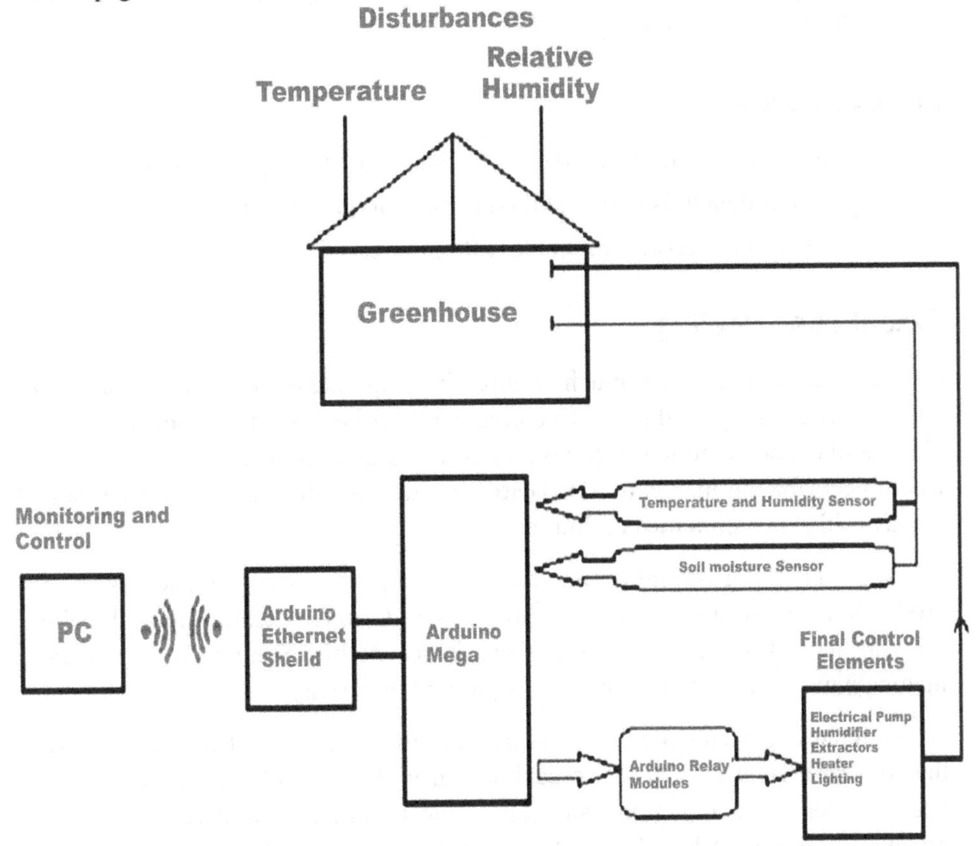

Use of computers for controlling greenhouse environment

The advantages and disadvantages of automated computer system control system are as under:

Advantages

1. Properly and appropriately equipped computer system with software can manage various programmes swiftly and coordinate with the devices meant to create optimal environment inside the greenhouse structures.
2. The data records of greenhouse structures during cropping span can be stored and analysed precisely to make any decision on management practices for climate control.

3. A centralized computer operating system can be used to monitor and control greenhouses at different locations from a single window which helps to use the resources with greater precision at lower costs.
4. Well designed and equipped automated system can make minor adjustments in anticipation of changes in weather for operating the system with desired results and energy saving.

Disadvantages

1. Requires high initial investment as well as demands high maintenance cost.
2. Automation demands highly skilled persons to handle the full process.
3. Exclusively very costly for small scale growers.

Greenhouse Heating

Greenhouse heating is very much required in temperate regions where the energy trapped inside the greenhouse structures may not be adequate especially during night hours. The some northern regions of India do experience extreme winters, where need for the installation of heating system is felt to cultivate high valued crops during such climatic conditions.

A good heating system is one of the most important steps for successful crop production. Any heating system that provides uniform temperature control without releasing harmful materials to the plants is acceptable. Suitable energy sources include natural gas, LP gas, fuel oil, wood and electricity.

Greenhouse heater requirements depend upon the amount of heat loss from the structure. Heat loss from a greenhouse usually occurs by all three modes of heat transfer *i.e.* conduction, convection and radiation. The heat demand of a greenhouse is normally calculated by combining all three losses as a coefficient of heat loss equation.

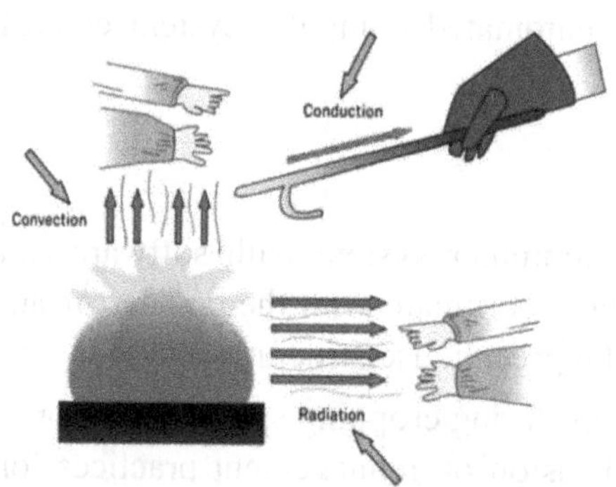

General concept of energy transfer

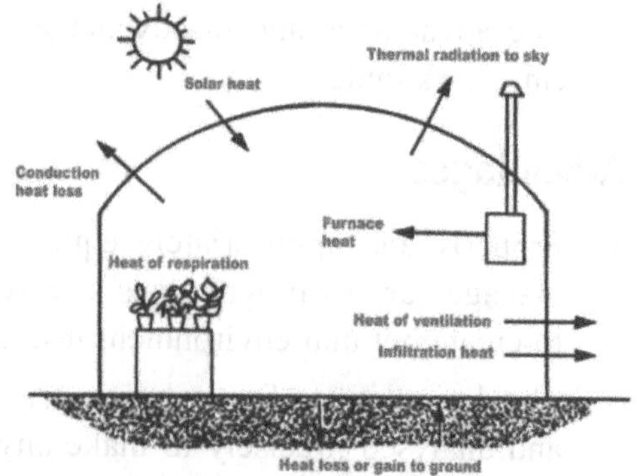

Modes of heat loss in greenhouse

Conduction

Heat is conducted either through a substance or between objects by direct physical contact. The rate of conduction between two objects depends on the area, path length, temperature difference and physical properties of the substance such as density. Heat transfer by conduction is most easily reduced by replacing a material that conducts heat rapidly with a poor thermal conductor (insulator) or by placing an insulator in the heat flow path. An example of this would be replacing the metal handle of a kitchen pan with a wooden handle or insulating the metal handle by covering it with wood. Air is a very poor heat conductor and therefore a good heat insulator.

Convection

Convection heat transfer is the physical movement of a warm gas or liquid to a colder location. Heat losses by convection inside the greenhouse occur through ventilation and infiltration (fans and air leaks). Heat transfer by convection includes not only the movement of air but also the movement of water vapours. When water upon evaporation inside the greenhouse absorbs energy and when water vapours condense back to a liquid, it releases energy. So when water vapor condenses on the surface of an object, it releases energy to the outside environment.

Radiation

Radiation heat transfer occurs between two bodies without direct contact or requires medium such as air for heat transfer. Like light, heat radiation follows a straight line and is either reflected, transmitted or absorbed upon striking an object. Radiant energy must be absorbed to be converted into heat. All objects release heat in all directions in the form of radiant energy. The rate of radiation heat transfer varies with the area of an object, and temperature and surface characteristics of the two bodies involved. Radiant heat losses from an object can be reduced by surrounding the object with a highly reflective, opaque barrier. Such a barrier (1) reflects the radiant energy back to its source (2) absorbs very little radiation so it does not heat up and re-radiates energy to outside objects (3) prevents objects from 'seeing' each other, a necessary element for radiant energy exchange to occur.

Factors Affecting Heat Loss

Heat loss by air infiltration depends on the age, condition and type of greenhouse. Older greenhouses or those in poor condition generally have cracks around the doors or holes in cladding material through which large amounts of cold air may enter. Greenhouses covered with large sheets of glazing materials, large sheets of fibre glass or a single or double layer of rigid or flexible plastic have less infiltration. The greenhouse ventilation system also has a large effect on infiltration. Inlet and outlet fan shutters often allow a large air exchange if they do not close tightly due to poor design, dirt, damage or lack of lubrication. Window vents seal better than inlet shutters, but even they require maintenance to ensure a tight seal when closed.

Solar radiation enters a greenhouse and is absorbed by plants, soil and greenhouse fixtures. The warm objects then re-radiate this energy outward. The amount of radiant heat loss depends on the type of glazing, ambient temperature and amount of cloud cover. Rigid plastic and glass materials exhibit the 'greenhouse effect" because they allow less than 4% of the thermal radiation to go back through the covering materials to the outside.

HEATING AND AIR CIRCULATION SYSTEMS

Central Heating

Central heating systems generally use hot water or steam distributed to heat exchangers made up of standard black pipe (natural convection), finned-pipe (natural convection), or steam/hot-water unit heaters (forced convection). Hot water may also be circulated through hollow structural members.

Greenhouses of 9 m or less in width may be heated with standard black pipe or finned-pipe placed only along the side walls. For single-span houses wider than 9 m, distribute the pipes along the side walls and below/between the crop (or under the benches) in proportion to the expected heat loss for the walls and roof, respectively. If steam is used as the heating medium, heating surfaces should be installed at least 0.30 m away from any plant material. For narrow-span ridge and furrow houses, heating pipes should be placed along the exterior walls and below the gutters between sections. Pipes may also be placed below benches, between the rows or adjacent to root media for root zone heating.

When using black or finned pipe, naturally induced air circulation does not always provide sufficiently uniform air temperatures at plant height. If increased temperature uniformity is desired, air circulation fans can be installed in accordance. Since maximum heating capacity is rarely needed all the time, central heating systems should be designed with two or three zones. The primary zone should be floor or under bench heat.

All boilers and hot water heaters shall have appropriate valves, controls and safety devices and shall be installed in accordance with all applicable codes. If heating pipes are galvanized or painted with aluminized paint, heat delivery rates will be approximately 15% less than from black pipe.

Unit Heaters

Unit heaters are available in steam/hot-water, direct oil fired and direct gas fired versions. Vertical discharge unit heaters placed overhead are not recommended because they may damage crop with the high temperature discharge air and they do not maintain sufficiently uniform temperatures in the crop canopy. For short statured crops on benches or raised beds, horizontal discharge unit heaters designed specifically for greenhouse heating can be used without additional circulation fans; however temperature variations within the greenhouse may exceed those in houses with black or finned-pipe heat exchangers. Horizontal air circulation or perforated tube air distribution can reduce these variations significantly.

The fans on the unit heaters may be operated continuously to provide improved air circulation. Two heaters with continuously operating circulating fans can be used in houses 20 m or less in length. Heaters should be located in diagonally opposite corners discharging toward the far walls in a direction parallel to the sidewalls. For greenhouses longer than 20 m, but less than 40 m, heater fans alone may not be sufficient to maintain air circulation patterns. In this case, it is recommended that two additional fans be placed at the middle of the greenhouse, one on each side with the air discharge directed toward the end walls, to aid air circulation.

Direct fired heaters shall be provided with outside air for combustion. It is recommended that this should be supplied through ducts with cross sections of 880 mm^2/kW of heater capacity. All gas or oil-fired heaters shall be vented in accordance with manufacturer's recommendations using stacks configured.

Perforated Tube Air Circulation

Air circulation is essential for minimizing CO_2, temperature, and moisture gradients in closed greenhouses. Overhead perforated plastic tube systems can be used for this purpose, although at reduced efficiency compared to horizontal airflow. The tube is generally connected directly to the outlet of a fan which forces air down the length of the tube and out through appropriately spaced discharge holes. If the system is used to distribute heat, unit heaters are typically installed at the inlet end of the tube system and the exhaust air of the heaters is directed into the inlet of the fan. The warm air discharge from each unit heater should be directed behind the plastic-tube air circulation fan. The capacity of the circulation fan shall be equal to or greater than the heater fan capacity. Air circulation fans should be run continuously during the heating season.

Horizontal Air Circulation

Horizontal air circulation consists of air circulation in a horizontal fashion above and within the plant canopy using large diameter, low power propeller fans. Fans should be located over the crop in each section. In single span greenhouses, fans should be installed with their axes parallel to the long dimension of the house at a distance of 1/4 the house width from the side walls. The air should be circulated parallel to the side walls down along one side and back along the other. Fan mounts should be designed to allow adjustment of the fans to insure that local velocities in the vicinity of the canopy do not exceed 1.0 m/s. Fans with shrouds that produce a longer throw distance are preferred.

Bench Heating

Bench heating can be used to provide optimum temperatures in the root zone for seed germination, propagation and general plant growth. It can also be used to reduce heating costs if air temperatures can be lowered to compensate for higher root zone temperatures. Hot water at temperatures of 35 to 40°C is generally circulated through 13 mm polyethylene, PVC, Chlorinated polyvinyl chloride (CPVC), polybutylene pipe or 6 mm EPDM (ethylene propylene diene monomer) tubing for maximum temperature uniformity. In this system, each heating loop has the same length and the water has to travel the same distance through the supply and return system to reach each heating loop. The design flow rate should be sufficiently high so that the temperature difference between the supply and return side in each loop remains small and potential sedimentation is avoided. On the other hand, the flow rate should not be so high as to increase friction losses. The pipe or tube spacing should be approximately 100 mm to allow the higher temperature on the supply side to be balanced by the lower temperature on the return side.

The pipe or tubing loops can be placed directly on the bench or attached underneath. Polystyrene insulation board placed immediately below the pipe or tubing will insure that most of the heat is directed into the root zone. Bench heating should be used as the first stage of heating.

Floor Heating

Floor heating is considered good practice where plant containers can be set directly on the floor. It is particularly well suited where a minimum amount of protection is needed as in over wintering greenhouse houses. Loose gravel, porous concrete (concrete without sand), solid concrete or sand can be used for the floor material. A floor thickness of 75 to 100 mm is generally recommended. If solid concrete is used, the floor should be sloped for drainage. Sand floors should be covered with perforated plastic to retain moisture (to promote heat transfer between the pipes and the sand). If ground water is within 1.8 m of the floor, 25 to 50 mm extruded polystyrene insulation board should be installed below the floor.

A typical installation circulates 35 to 40 °C water at velocities of 0.61 to 0.91 m/s through 20 mm polyethylene, PVC or polybutylene pipe embedded in the floor. The heating pipes should be configured with a 'reverse return' header system with supply and return headers at the same end of the greenhouse. A typical floor heating system provides 47 to 78 W/m^2 heat depending on the crop density on the floor and the desired temperature difference between the root zone and the greenhouse air.

Heating Controls (Use of sensors)

Heating controls and sensors should be capable of withstanding extreme humidity and dust conditions. Sensors should be continuously aspirated at a speed of 3.0 to 5.1 m/s to reduce temperature gradients in the vicinity of the sensors. Aspirator fans should have totally enclosed motors and should pull the air across the sensors to avoid temperature measurement errors due to the addition of motor heat to the airstream. The control sensors should be fully shaded from direct and diffuse solar radiation. The material used to shade the sensors should have a low thermal conductivity and should be painted or coated white on the side facing external radiation to minimize energy absorption.

Sensors should be located near the centre of the growing area. The location should be representative of that occupied by the plants and should not be unduly affected by heating and ventilating systems or structural members.

Heat Conservation Practices

Energy can be saved significantly if a grower implements the following suggestions of the American Council for Agricultural Science and Technology:

1. Use fiberglass or polyethylene on the interior gable ends.
2. Close all the possible openings.
3. Keep steam or hot water system well maintained to prevent leakage.
4. Use reflector materials behind heating pipes to reflect heat into the greenhouse.

5. Install and maintain the automatic valves in the heating system.
6. Insulate hot water, steam supply, return piping and inspect them at regular intervals, replace the insulation according to requirement.
7. Maintain the boiler at peak efficiency.
8. Monitor the thermostats regularly for proper operation.

Other Improved Practices for Energy Conservation

1. Covering a greenhouse with a double layer of polyethylene to reduce the loss of heat energy.
2. Placing a removable sheet of polyethylene over the crop and a row cover over each plant row in order to reduce heat loss from the greenhouse during night.
3. Application of opaque sheets as curtains during night.
4. Application of at least one layer of movable thermal screen.
5. Installation of polyethylene tubing which, when installed, seals the growing area from the roof surface area.

16

GREENHOUSE: COMPONENTS, ACCESSORIES AND BIS CODES

Components of Greenhouse

Roof: transparent cover of a greenhouse.

Gable: transparent wall of a greenhouse

Cladding material: transparent material mounted on the walls and roof of a greenhouse.

Rigid cladding material: cladding material with such a degree of rigidity that any deformation of the structure may result in damage to it. *e.g.* Glass

Flexible cladding material: cladding material with such a degree of flexibility that any deformation of the structure will not result in damage to it. *e.g.* Plastic film

Gutter: collects and drains rain water and snow which is place at an elevated level between two spans.

Column: vertical structure member carrying the green house structure

Purlin: a member which connects cladding supporting bars to the columns

Ridge: highest horizontal section in top of the roof

Girder: horizontal structure member, connecting columns on gutter height

Bracings: to support the structure against wind

Arches: Member supporting covering materials

Foundation pipe: connection between the structure and ground

Span width: centre to centre distance of the gutters in multispan houses

Greenhouse length: dimension of the greenhouse in the direction of gable

Greenhouse width: dimension of the greenhouse in the direction of the gutter

The structure of a typical greenhouse with its components is shown in **Fig. 16.1**.

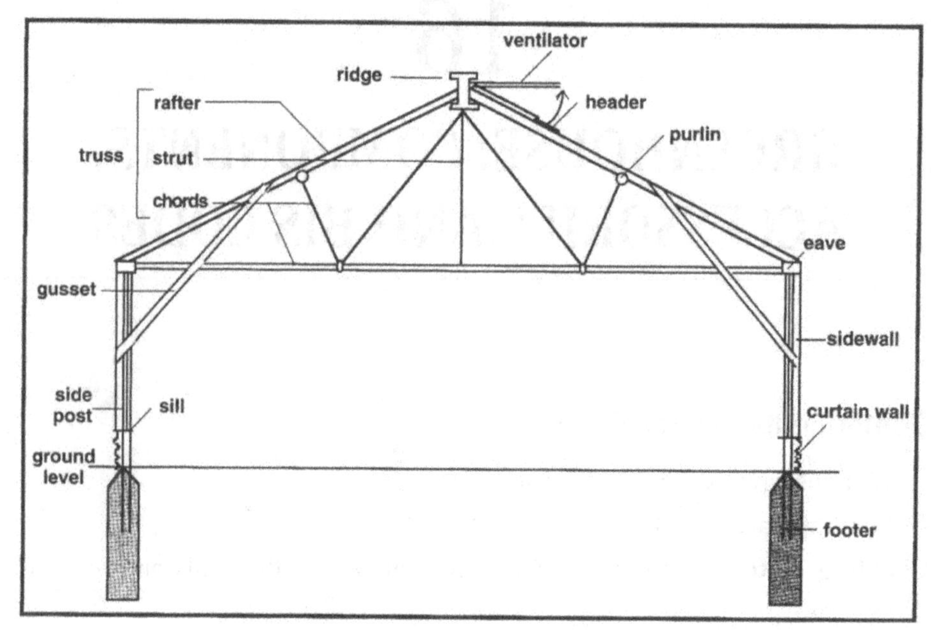

Fig. 16.1: Components of Greenhouse

Greenhouse Accessories:

System type	Accessories
Climate Controls	Heating and Cooling, Thermostats, Variable Speed Controls, Humidistats, Cycle Timers
Ventilation	Evaporative Coolers, Exhaust Fans & Shutters, Automatic Vent Openers, Circulation Fans
Heaters	Gas Heaters, Electric Heaters, Heating Mats & Cables
Misting Systems	Sprinkler System, Misting Systems, Mist Timers & Valves, Water Filter
Watering	Plant Watering Systems, Drip Systems, Professional Water Hoses, Watering Timers, Overhead Watering Systems
Meters	Min./Max. Hygro-Thermometer, Min/Max Thermometer, Soil Thermometer, Thermo-Hygrometer, Light Intensity Meter, pH Meter, EC Meter, LUX meter
Greenhouse Plastics	Shade Fabric, Greenhouse Film, Patching Tape, Batten Tape, Ground Cover Flooring, Greenhouse Bubble Insulation
Grow Lights	CFL Fluorescent Lights, T5 Fluorescent Light Systems, MH & HPS Light Systems, MH & HPS Bulbs & Supplies
Benches and Shelves	Superior Greenhouse Benches

[Table Contd.

Contd. Table]

System type	Accessories
PVC Fittings and Pipe	PVC Fittings, PVC Pipe
Drip line fittings	Grommet, Elbow, Nipple, Joiner, Reducing Tee, Tee, Reducer, End Cap, Lateral Cock
Greenhouse Electrical	Waterproof Outlets, Electrical Wire, Flexible Conduit
Pollination	Electric Pollinator/ Pollination Tool

Equipments Required for Drip Irrigation System

- A pump unit to generate 2.8 kg/cm^2 pressure
- Water filtration system (sand/silica/screen filters)
- PVC tubing with dripper or emitters

Types of Drippers

- **Labyrinth drippers:** A **labyrinth dripper** features a wide passage with indentations. These indentations bring swirls in the water flow that will cause the pressure to drop and keep the water moving.
- **Turbo drippers:** These drippers have three different units such as base, cap and a disk. The base of dripper has a small inlet, which is inserted on into a drip lateral or hose, and the cap has an outlet from where the water is discharged.
- **Pressure compensating (PC) drippers:** PC is a term used to describe an emitter that maintains the same output at varying water inlet pressures. PC emitters deliver a precise amount of water regardless of changes in pressure due to long rows or changes in terrain. There is a flexible diaphragm inside the dripper that regulates the water flow and tends to flush particles from the system (self-flushing). These drippers contain silicon membrane which assures uniform flow rate
- **Button drippers:** Button drippers are appropriate for most drip watering applications on a flat terrain. They are easy and simple to clean.

Depending upon the type of water, different kinds of filters can be used

Gravel filter: It is used to filter water obtained from open canals and reservoirs contaminated by organic impurities like algae. The filtering is done by beds of basalt or quartz.

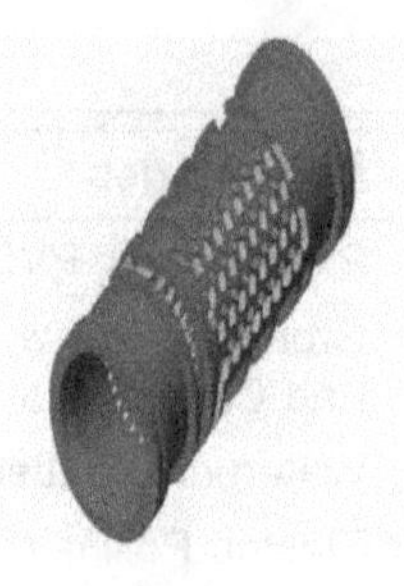
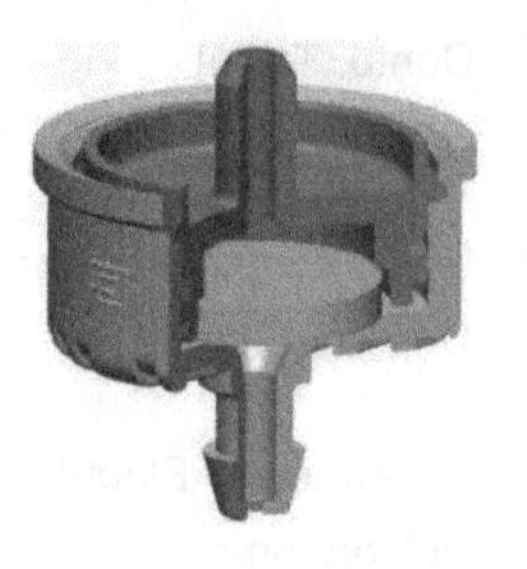

On line dripper **On line dripper-PC type** **In line Dripper** **No Drain, PC Dripper**

Hydrocyclone: It is used to filter well or river water that carries sand particles.

Disc filters: It is used to remove fine particles suspended in water.

Screen filters: It consists of stainless steel screen of 120 mesh (0.13mm) size and is used for second stage filtration of irrigation water.

Sand Filter **Hydrocyclone Filter** **Screen Filter** **Disc Filter**

Equipments for Fertigation

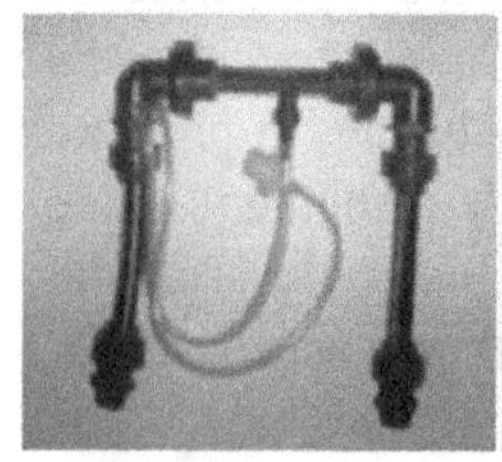

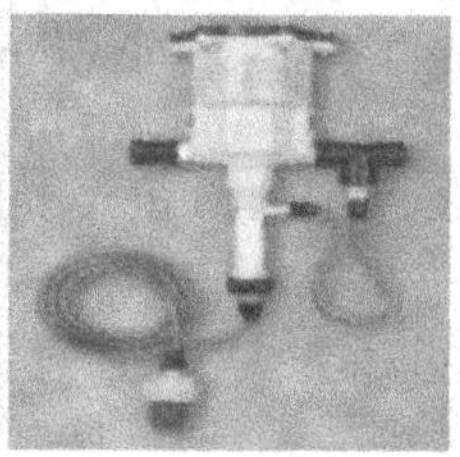

Venturi Injector **Fertilizer tank** **Injector pump** **Automatic Fertigation Controller**

BIS Standards for Protected Cultivation:

Applications	Component Description	BIS Code
Mulching	Surface covered cultivation-plastics mulching-code of practice	IS 15177:2002
Greenhouse	Plastic film for Greenhouses -specifications	IS 15827:2009

[Table Contd.

Contd. Table]

Applications	Component Description	BIS Code
	Recommendations for Layout, Design and Construction of Greenhouse Structures.	IS 14462 : 1997
	Recommendation for Heating, Ventilating and cooling of Greenhouses	IS 14485 : 1998
	Steel Tubes for Structural Purposes	IS 1161 : 1998
Agro Shade nets	Shade-nets for Agriculture and Horticulture Purpose	IS 16008:2012
Protection Nets	Plant protection nets	IS 10106: part 1: section 6:1992

BIS Standards for Micro-irrigation System (MIS)

Sr. No.	Component Description	BIS Code
1.	Polyethylene pipes for Irrigation- Laterals with amendment number 5	IS 12786: 1989 (reaffirmed 1998)
2.	Emitters	IS 13487: 1992
3.	Emitting pipes system	IS 13488: 2008
4.	Strainer type filters	IS 12785: 1994
5.	Irrigation equipment rotating sprinkler Part II, Test method for uniformity of distribution (1st revision) (amendment 1) (Including Rain gun)	IS 12232 (Part II) 1995
6.	Polyethylene microtubes for drip irrigationsystem	IS 14482: 1997
7.	Fertilizer and Chemicals Injection system Part I Venturi Injector	IS 14483 (Part 1) 1997
8.	Micro Sprayers	IS 14605: 1998
9.	Media Filters	IS 14606: 1998
10.	Hydro Cyclone separators	IS 14743: 1999
11.	PVC pipes for water supply	IS 4985 – 1999
12.	Irrigation equipment sprinkler pipes specifications Part I Polyethylene pipes	IS l4151 (Part I) 1999
13.	Irrigation equipment sprinkler pipes specifications Part II Quick couples Polyethylene pipes	IS l4151 (Part II) 1999
14.	Quality of Irrigation water	IS 11624 : 1986
15.	HDPE Pipes	IS 4984 : 1995
16.	Moulded PVC Fittings	IS 7834 : 1987
17.	GI and MS Fittings	IS 1879 : 1987

[Table Contd.

Contd. Table]

Sr. No.	Component Description	BIS Code
18.	GM Valves	IS 778 : 1984
19.	CI Non Return Valves	IS 778 : 1984
20.	Fabricated PVC Fittings	IS 10124 : 1988
21.	GI Pipes	IS 1879 : 1987
22.	Sluice Valves	IS 780: 1984
23.	PE Fabricated Fittings	IS 8360: 1977
24.	PE Moulded Fittings	IS 808: 2003
25.	PVC Foot Valves and NRV	IS 10805: 1986
26.	Irrigation equipment rotating sprinkler Part I, Design and Operational requirements (1st revision)	IS 12232 (Part I) 1996
27.	Design, Installation and Field evaluation of MIS	IS 10799: 1999
28.	Prevention and treatment of blockages problems in drip irrigation systems	IS 14791: 2000

17

IRRIGATION MANAGEMENT IN GREENHOUSE PRODUCTION SYSTEM

What is Irrigation?

It is an essential agricultural practice that provides desired quantity of water to crop at optimal intervals.

Role of Irrigation Water in Agriculture System

- Sustains soil biological and chemical activity and mineralization during dry periods.
- Promotes soil solution and nutrient uptake.
- Promotes carbohydrate production through the process of photosynthesis.
- Water molecules present within the water conducting vascular bundles and other tissues of plants provide physical support to the plant.
- Maintains optimal temperatures within the plant through the process of evapo-transpiration.
- Protects crops from frost damage by increasing temperature in orchard systems during the threats of frost damage.
- Reduces plant stress and improves crop quality.

Quality Criteria for Irrigation Water

Water plays an important role in photosynthesis process and its quality is determined by the following characteristics:

- Salinity hazard: Total soluble salt content
- Sodium hazard: Ratio of sodium (Na^+) to calcium (Ca^{2+}) and magnesium (Mg^{2+}) ions

- pH of irrigation water
- Alkalinity: Carbonate and bicarbonate; specific ions: chloride (Cl^-), sulphate (SO_4^{2-}), boron (BO_3^{3-}) and nitrate-nitrogen (NO_3-N)
- Organic contaminates
- Oil pollutants and heavy metals

Greenhouse Irrigation System

An irrigation system must be designed to apply precise amount of water to satisfy crop requirements each day throughout the year. The water requirement of plants depends on the crop type, region, growing season, weather conditions and the operating efficiency of heating or ventilation system inside the greenhouse. Water needs varies with medium and is also dependent on soil type, soil mix, size and type of the container or bed. Avoid excessive or deficient irrigation as it is injurious to the crop. Several irrigation systems such as hand watering, perimeter watering, overhead sprinklers, boom watering and drip irrigation are currently in use in greenhouse production system.

Rules of Watering

There are three important rules of water application:

1. **Use a well drained substrate with good structure:** A well drained and aerated substrate is necessary to achieve proper watering. Therefore, substrate with high moisture retention capacity along with good aeration is essential for proper growth and development of plants. Substrates having such characteristics can not only be obtained from field soil but can be formulated by mixing desired combination of coarse texture and highly stable structure.

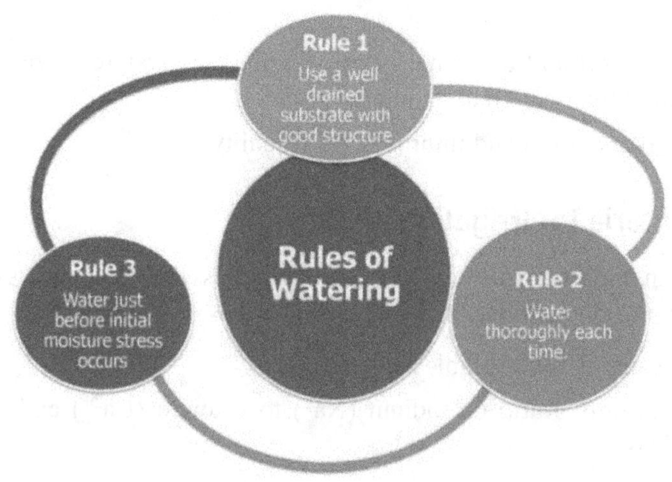

2. **Water thoroughly each time:** Avoid partial watering of the substrates. Make sure that the supplied water should flow from the bottom in case of containers and the root zone should be uniformly wetted in case of beds. In general, 10 to 15% excess water is supplied than the requirement of plant. The water requirement for soil based substrates is at a rate of 20 L/m^2 of bench and 0.3 to 0.35 litres per 16.5 cm (6.5") diameter pot.
3. **Water just before initial moisture stress occurs:** Irrigate the plant just before the plant starts showing early symptoms of water stress. Avoid excess watering as it reduces the aeration and root development. The symptoms of water stress differ with crop. It can be identified by several foliar symptoms like texture, colour and turbidity. The crops which do not show any symptoms of water stress, can be acquired through colour, feel and weight of the substrate.

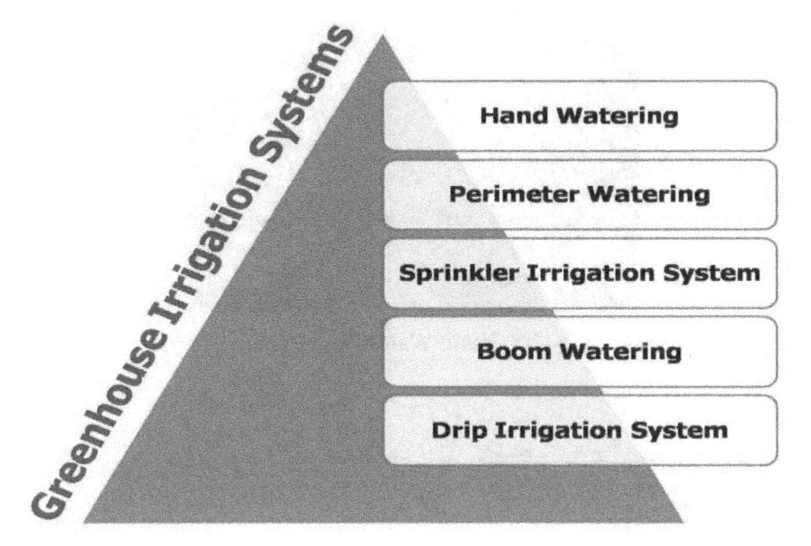

Hand Watering

Hand watering is the most traditional method of irrigation but now-a-days it is uneconomical and time consuming. Hand watering is affordable in such situations where plant density is high as in case of seed beds, pots or an area that dries sooner and requires frequent irrigation. Moreover, there is immense risk of applying too little water or long interval between two subsequent watering. Hand watering is very time consuming and tiresome practice. It can easily performed by inexperienced and unskilled persons. Automatic watering is quick and easy method which can be performed by a single person or the grower himself. While practicing hand watering, a water breaker should be placed on the end of the hose to break the force of the water, permitting a higher flow rate without washing the root substrate out of the bench or pot. It reduces the risk of structural damage of the substrate surface.

Perimeter Watering

Perimeter watering is a typical system consisting of a polyethylene or PVC pipe around the perimeter of a bench with nozzles. It can be used for crop production in benches or beds as it spray water over the substrate surface below the foliage without striking the foliage.

Hand Watering

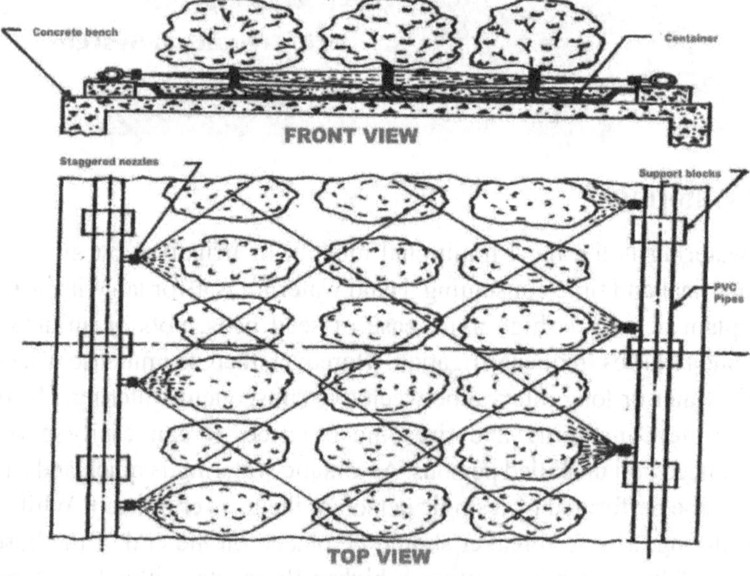

Perimeter Watering

PVC pipe has the advantage of being very stationery, pipe tends to roll if it is not anchored firmly to the side of the bench. It supports nozzles to rise or fall from proper orientation to the substrate surface. Nozzles made up of nylon or a hard plastic are used to put out a spray arc of 180°, 90° or 45°. They are arranged across the benches in such a way so that each nozzle projects out between two other nozzles on the opposite side. Perimeter watering systems with 180° nozzles require one water valve for benches up to 30.5 m in length. But, while using alternate nozzles of 180° and 90° or 45°, the length of a bench should not exceed 23 m for service of one water valve. For benches over 30.5 m and upto 60.1 m, water pipes should be installed on either side, one to serve each half of the bench. This system applies 1.25 L/min/m of pipe.

Overhead Sprinklers

Majority of crops cannot tolerate wet leaf surface and become infected with fungal or bacterial diseases. Such crops require dry foliage to avoid disease infection. But, few bedding plants like azalea liners and some green foliage plants can survive without any trouble in such condition and can easily be irrigated from overhead sprinklers. A pipe attached with riser pipes just above the height of a crop is installed at periodic space along the bed. A nozzle is installed at the top of each riser. About 0.6 m height is sufficient for bedding plants and 1.8 m for fresh flowers. Nozzles vary from those that throw a 360° pattern continuously to types that rotate around a 360° circle. Sometimes, trays are placed under pots to avoid waste of water that fall on the ground between pots. The trays also contain drainage holes for removal of excess water and the stored quantity of water is subsequently absorbed by the substrate.

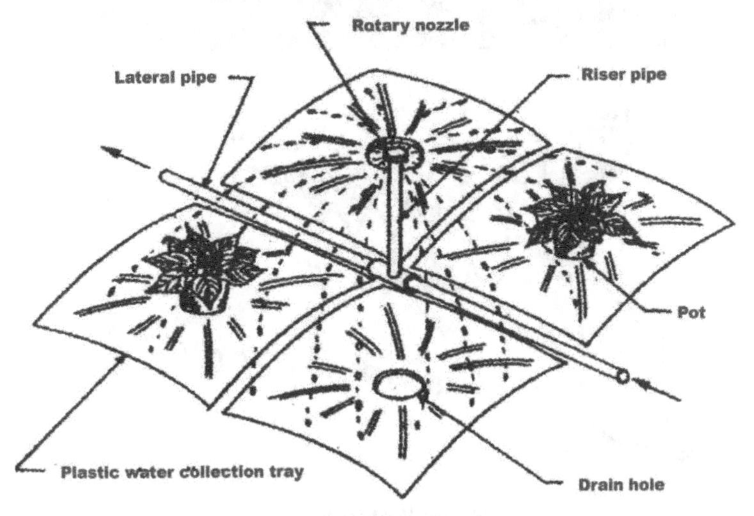

Overhead Sprinkler watering system

Boom Watering

Boom watering is most frequently used for raising plug tray seedling. It can function either as open or a closed system. Plug trays containing 100 to 800 cells with dimensions of 30 cm × 61 cm (length x width) and depth of 1.3 to 3.8 cm are used in plug tray seedlings. Each seedling is grown in its individual cell. Accuracy of watering is very important during production of plug seedlings.

A boom watering system generally consists of a water pipe boom that extends from one side of a greenhouse bay to the other. The pipe is fitted with nozzles that can spray either water or fertilizer solution down onto the growing seedlings. The boom is attached at its center point to a carriage that rides along rails, often suspended above the centre walk of the greenhouse bay. In this way, the boom can pass from one end of the bay to the other. The boom is propelled by an electric motor. The quantity of water delivered per unit area of plants is adjusted by the speed at which the boom travels.

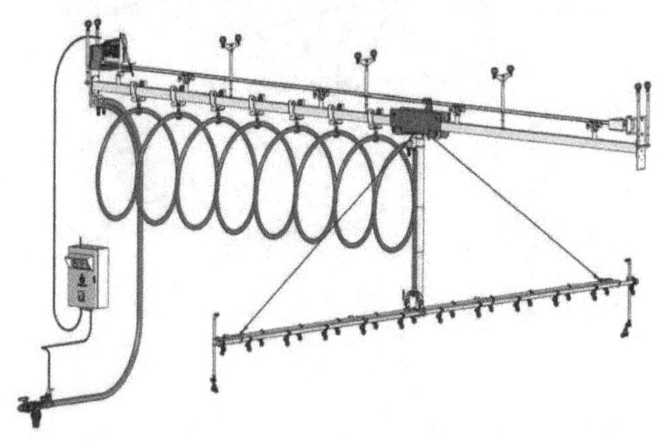

Boom Watering

Drip Irrigation

Drip irrigation, commonly known as trickle irrigation, consists of plastic tubes of small diameter laying on the surface or subsurface of the greenhouse or field beneath the plants. Small holes or emitters present along the tube at frequent intervals delivers water to the plant. Drip irrigation in combination with protected agriculture is commonly used as an essential part of the comprehensive design. It is the only means of applying uniform water and fertilizer to the plants in row covers and greenhouse. Drip irrigation assures optimum production with minimal use of water which results in reduced cost of water, fertilizer, labour and machinery. Drip irrigation is the best means of water conservation and provides maximum control over environmental variability. The water application efficiency of this system is 90 to 95% as compared to sprinkler and furrow irrigation. Drip irrigation is recommended for crop production in protected structure as well as open field in arid and semi-arid regions of the world.

Drip irrigation is replacing surface irrigation where water is scares or expensive, when the soil is too porous or too impervious for gravity irrigation, land leveling is impossible or very costly, water quality is poor, the climate is too windy for sprinkler irrigation, and where trained irrigation labour is not available or is expensive. Drip irrigation reduces weed growth as the irrigation water is applied directly to the plant row and not to the entire field like sprinkler, furrow or flood irrigation. In this system, fertilizer is directly injected to the root zone of the plant with irrigation water which increases the fertilizer use efficiency. Plant foliage diseases may be reduced since the foliage is not wetted during irrigation.

High initial installation cost is one of the major disadvantages of drip irrigation as compared to other irrigation systems. However, these costs must be evaluated through comparison with the expense of land preparation and maintenance often required by surface irrigation.

The head between the pump and the pipeline network usually consists of control valves, couplings, filters, time clocks, fertilizer injectors, pressure regulators, flow meters, and gauges. Since the water passes through very small outlets in emitters, it is an absolute necessity that it should be screened, filtered, or both before it is distributed in the pipe system. The initial field positioning and layout of a drip system is influenced by the topography of the land and the cost of various system configurations. Design considerations should also include the relationship between the various system components and the farm equipments required to plant, cultivate, maintain and harvest the crop.

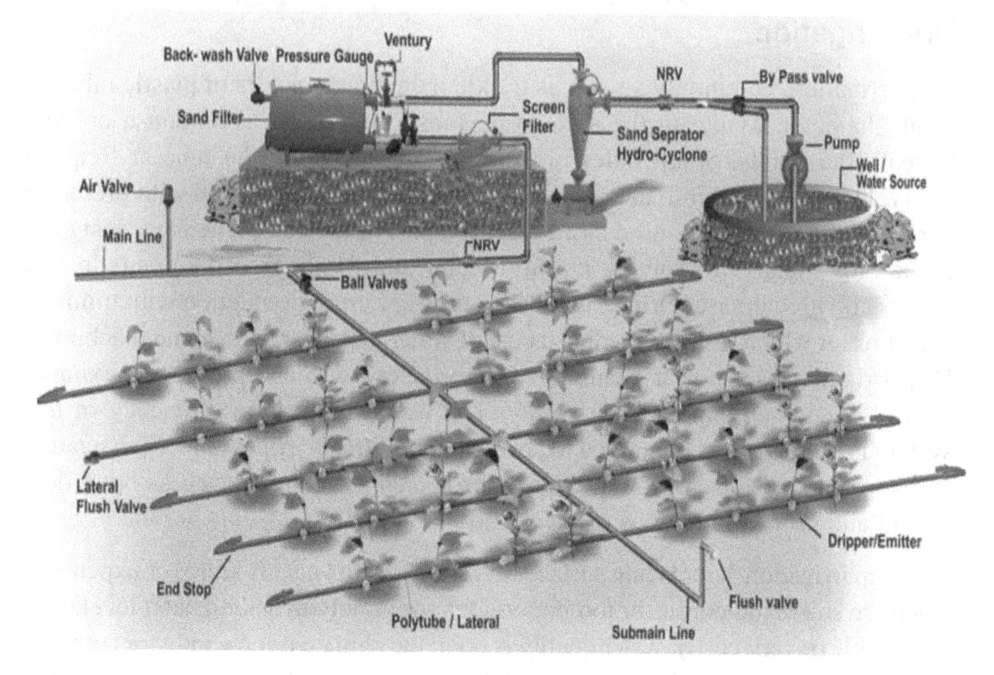

Typical layout of drip irrigation system

Courtesy: Jain Irrigation

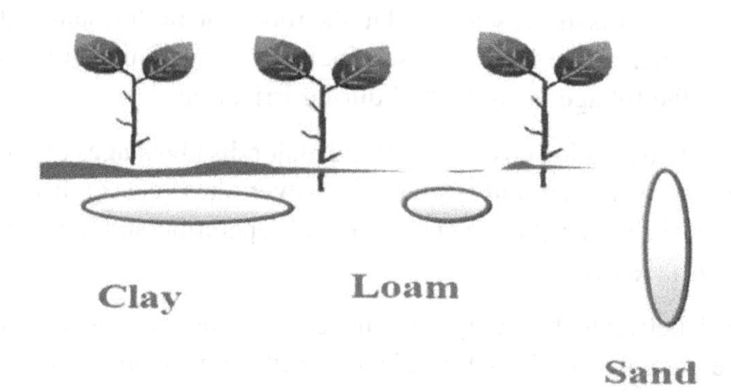

Water distribution pattern in different soil types

Mat Irrigation

Mat irrigation system consists of a specialized mat with water supplied by drip lines which remain moist and supply constant water to plants. The pots placed on the mat wicks up moisture from the mat and the soil through drainage holes. As the soil loses water due to evaporation from the soil surface, the soil continuously absorbs water from the mat through capillary action until it is evenly moist. This system avoids risk of over or under watering to plants.

Drip Tubing

Drip tubes supply water directly to root system of plant. Multiple tubes attached to a water supply extend outward so that each line provides supply to one pot. An emitter provides a small spray of water onto the top soil. As the system supply water through properly timed drip tube, it prevents water from striking the foliage and avoids risk of bacterial infections from standing water droplets. The system can be controlled manually or automatically through timers and moisture sensors.

Overhead Misters

In this irrigation system, plants are irrigated with sprinkler heads connected to overhead water pipes which emit a misty spray across the entire greenhouse and cover all the pots evenly with moisture. This system is beneficial for large greenhouses that are having same plant species because the sprinklers cover a significant large area in a short period of time. One major drawback of this irrigation method is wasting water as moisture finds its way onto paths and other areas rather than plants.

18

GROWING MEDIA FOR GREENHOUSE CULTIVATION

The cultivation of crops under greenhouse production system is based on two types of growing media: (1) Soil-based growing media and (2) Soil-less media

Soil Based Growing Media

Most of the greenhouse growers prefer soil based growing media. It is an easiest way to start crop production in greenhouse. Usually, the soil medium is prepared by mixing equal proportion of loam field soil, coarse sand and well decomposed organic matter on volume basis. The pH of media is generally adjusted to 6.0-6.5. Most of the greenhouses use red laterite soils for soil based greenhouse cultivation as the range of pH and EC of such soil is highly suitable for growing plants. The soil with pH in this range makes easy availability of nutrients to plants.

The growing media in polyhouse should proportionally compose of 70% red laterite soil, 20% FYM and 10% rice husk/sand. The proportion of organic matter should be kept more while using sandy soils as growing media but in case of clay soil, the proportion of sand should be more. Sometimes, coarse organic materials like coconut husk, coarse bark, peat moss *etc.* are also replaced with parts of well decomposed organic matter to make the growing medium more porous. Such kind of adjustment in composition of growing media may improve several characteristics like water holding capacity, aeration, steady nutrient supply *etc.*

Maintenance of soil-based media

Soil based media losses organic matter due to decomposition, so organic matter should be applied periodically. Organic matter at the rate of 5 to 10% of the volume of growing media depending upon the nutrient and drainage status proves to be good.

However, soil based media is associated with some disadvantages and these are:

1. Soil-based media contains various weed seeds, pathogens and insect-pests which necessitates pasteurization of media at regular intervals.
2. Inappropriate combination and proportion in soil media cause problem of aeration and drainage.

Management of Soil pH

- **Lime [Calcium carbonate ($CaCO_3$)] is used to raise the soil pH.**
- **Gypsum [Calcium sulfate ($CaSO_4$)] is used to lower the soil pH.**
- **Each tonne of Gypsum or lime will bring about 0.1 change in the soil pH.**

Management of Saline Alkali Soils

The presence of exchangeable sodium in these soils is very high, which make such soil almost impervious to water. So, application of gypsum is required in large quantity to replace sodium as well as to make this element unavailable to the plants by leaching process. Gypsum @ 0.25 to 0.50 tonne per 1000 m^2 is applied and mixed in the soil by harrowing 2 to 4 weeks prior to sowing/planting of crops. Some fertilizers (ammonium sulfate, urea, and ammonium nitrate) when applied in the soil make an acid reaction, thus help in lowering down the pH or maintaining pH. Certain organic materials acidic in reaction such as pine needles or peat moss can also lower soil pH but at gradual rate over many years. Rapid reclamation of alkaline soils can be achieved with the application of sulphur, because it forms sulphuric acid which speeds up the process of lowering the PH. Still, the process of lowering the pH is slow and may take 1 to 2 years to get desired results. **Table 18.1** shows the quantity of sulphur required to lower the pH to a level of 6.5, while **Table 18.2** depicts the details of commonly used materials for lowering down the pH and their equivalent amendment values..

Table 18.1: Requirement of sulphur to lower the soil pH to 6.5

Original pH	Sulphur quantity (Tonne/acre)	
	Sandy Soil	Clay Soil
8.5	0.07-0.10	0.10-0.13
8.0	0.05-0.07	0.07-0.11
7.5	0.02-0.03	0.04-0.05

Table 18.2: Commonly used materials and their equivalent amendment values

Material (100% Basis)*	Chemical Formula	Tonnes of Amendment Equivalent to 1 Tonne of Pure Gypsum	Tonnes of Amendment Equivalent to 1 Tonne of Soil Sulfur
Gypsum	$CaSO_4.2H_2O$	1.00	5.38
Soil Sulphur	S	0.19	1.00
Sulphuric acid (Concentrated)	H_2SO_4	0.61	3.20
Ferric sulphate	$Fe_2(SO_4)_3\ 9H_2O$	1.09	5.85
Lime sulphur (22% S)	CaS	0.68	3.65
Calcium chloride	$CaC1_2\ H_2O$	0.86	–
Aluminum sulphate	$A1_2(SO_4)_3$	–	6.34

* The percent purity is given on the bag or identification tag

Gypsum

Chemically, gypsum is $CaSO_4.2H_2O$ and in its pure form, contains 18.6% sulphur (S) and 23% calcium (Ca). Hence, it is used in a variety of ways *viz.*, sulphur fertilizer, calcium fertilizer and soil amendment at application rates. Gypsum when used in smaller quantity as S or Ca fertilizers, has little effect on soil EC, however higher dose of gypsum when used for soil amelioration may increase the soil EC if the resultant soluble salts are not properly drained or leached out. The brief description on the mode of action of gypsum is given below:

Gypsum when applied for the amendment of sodic or magnesic soils, helps in displacement of Na and/ or Mg on the surface of clay particles with Ca, thus improves soil structure. Gypsum is not easily soluble, however the sodium and/ or magnesium) sulphates formed after its application are highly soluble. So, these soluble salts can easily be leached out with irrigation water, keeping them far away from the reach of growing plants. Though, the soil structure may get improved if the process of leaching of salts is not followed, but accumulation of salts in the soil may affect crop growth adversely high EC.

Checking of Soil EC (Electrical Conductivity) Prior to the Application of Gypsum in Sodic Soils

When soil analysis reports higher EC, application of gypsum is not recommended. So, it is better to improve soil drainage, as most of the sodium will be present as soluble sodium salts and not exchangeable sodium.

Management of acidic soils

Acidic soils can be ameliorated with liming materials like ground lime stone ($CaCO_3$), burnt lime (CaO) or hydrated lime $Ca(OH)_2$. Upon application of lime, soil solution becomes charged with calcium ions which in turn replace hydrogen ions in the exchange complex. When burnt / slaked /hydrated lime is applied in acidic soils, gets changed in calcium bicarbonate, which in solution reacts with soil colloids.

Requirement of Lime Requirement in Soil

1. pH or intensity of soil acidity: Approximately 0.15 tonne $CaCO_3$ per 1000 m^2 is needed to raise the pH by one unit (5.0 to 6.0).
2. Texture of the soil: Because of their higher buffer capacity, clay loam soils needs more lime than sandy soils.
3. Purity of the liming material.
4. Fineness of the material: Finer the material, more rapid is the effectiveness because of fast solubility. It has been observed that material which can pass through 60 mesh sieve easily, has more effectiveness
5. Chemical composition of the material: The neutralizing value of $CaCO_3$ is considered as 100.

Method of Lime Application

The application of lime should be ensured well in advance prior to sowing or planting of crop and it should be thoroughly mixed with soil for uniform results. Lime should not be applied in bulk rather its application should be targeted with small quantities every year or once in 2 years for getting more efficient results. Lime can applied in single dose, when the required quantity of lime is not more than 0.5 tonne per 1000 m^2 and if required quantity is more, then one half is applied before ploughing and the remaining half is applied and mixed after ploughing. Care must be taken mix lime properly in top 15-20 cm layer of soil and the quantity of lime required to raise pH from 4.5 to 5.5 and 5.5 to 6.5 (by 1.0 unit) is also dependent on type of soil **(Table 18.3)**. The list of common liming materials is given in **Table 18.4.**

Table 18.3: Requirement of finely ground limestone to raise soil pH (15-20 cm)

	Lime Requirements (Tonne per 1000 m^2)	
Soil Texture	**Raising pH from 4.5 to 5.5**	**Raising pH from 5.5 to 6.5**
Sand and loamy sand	0.05	0.06
Sandy loam	0.08	0.13
Loam	0.12	0.17
Silt loam	0.15	0.2
Clay loam	0.19	0.23

Table 18.4: List of common liming materials

Name of liming material	Chemical Formula	Equivalent $CaCO_3$ (%)	Source
Shell meal	$CaCO_3$	95	Natural shell deposits
Limestone	$CaCO_3$	100	Pure form, finely ground
Hydrated lime	$Ca(OH)_2$	120-135	Steam burned
Burned lime	CaO	150-175	Kiln burned
Dolomite	$CaCO_3$–$MgCO_3$	110	Natural deposit
Sugar beet lime	$CaCO_3$	80-90	Sugar beet by-product lime
Calcium silicate	$CaSiO_3$	60-80	Slag

Soil-less growing media

Different types of soil-less media are used for the cultivation of greenhouse crops and these are:

A. **Peat:** Peat is the most widely used soil-less medium because of its wider availability and relatively low cost. It consists of the remains of aquatic, marsh bog or swamp vegetation, which has been preserved under water in a partially decomposed state. Composition of peat varies widely, depending upon the state of decomposition, vegetation from which it is originated, nutrients and degree of acidity. It is very stable source of organic material that does not decompose quickly and holds a large quantity of water and air. Peat is quite acidic in nature (pH 3.5-4.0), therefore it is recommended to add limestone to adjust the pH. Mostly fresh and light coloured peat is used as it provides a better air space than older and darker peat.

Peat Perlite Vermiculite

B. **Perlite:** Pertile is grayish white coloured silicaceous material of volcanic origin mined from lava flows. The crude material is crushed, screened and heated at high temperature to remove moisture from the particles and explode the particle to small, sponge like kernels. The high processing temperature gives a sterile product. It is very light in weight only 2.7 to 3.6 kg /cubic feet. It is neutral in pH and has negligible cation exchange capacity. Perlite pieces create tiny air tunnels that allow free water and air movement to the roots. Water get adhered to the surface of perlite but it is not absorbed into the perlite aggregates. Although, costs are moderate but perlite is an effective amendment for growing media.

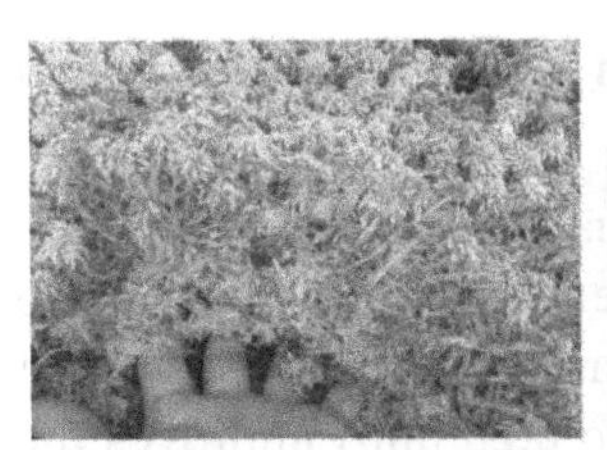

Sphagnum moss Leaf mold

C. **Vermiculite:** Chemically, vermiculite is a hydrated magnesium-aluminum-iron silicate which originates from micaceous minerals and get markedly expands when heated. It is very light in weight (2.7 to 4.5 kg/cubic ft.), insoluble in water and neutral in reaction. Vermiculite is very desirable component of soil-less media because of its high nutrient and water retention capacity, good aeration and low bulk density. It is generally not used with soil.

D. **Sphagnum moss:** Sphagnum moss is the hydrated remains of acid bog plants such as *Sphagnum pappilosum and Sohagnum palustre.* It is relatively sterile and light in weight. It have high water holding capacity as it consists of water holding cells which are able to absorb 10 to 20 times more water of its weight. It also consists of small amount of minerals.

E. **Leaf mould:** Generally leaf type such as maple, oak, sycamore and elm are used for leaf mold. Leaf mold is ready to use after the composting of leaves for 12 to 18 months. It should be made sterile before bringing into actual use as it may contain nematodes as well as weed seeds and noxious insects and diseases.

F. **Bark:** Bark is partially composted and screened material of plant origin. It is beneficial to obtain bark from local sources to avoid cost of transport. Bark contains varying quantities of cambium and young wood when it is removed from logs.

G. **Rockwool:** Rockwool is produced by burning a mixture of coke, limestone, basalt and possibly slog from iron production. The rockwool is composed of approximately 60% basalt, 20% coke and 20% limestone. It can be formulated in prescribed density according to the air and water holding requirement. It is available in various forms like cubes, slabs, granules *etc.* It is generally used in the form of cubes for propagation and in slabs for finishing crops. The granular form has very high available water and aeration properties. Insulation and acoustical grade rock wool is suitable for plant growth.

H. **Polystyrene foam:** It is white coloured synthetic product consisting of numerous closed cells filled with air. It helps in improving aeration and imparts light weight to rooting media. It is extremely light in weight, does not absorb water and has no appreciable CEC. It is neutral in reaction and thus, does not affect pH of rooting substrate. Polystyrene is available in the form of beads and flakes in various diameter from 3 mm to 10 mm for pot plant substrate. It is highly resistant to decomposition.

I. **Sand:** Sand is a basic component of soil and a valuable amendment for both potting mix and propagation media. Its particle size ranges from 0.05 mm to 2.0 mm in diameter. Medium and coarse sand particles provide optimum adjustment in media texture. Fine sand (0.05 mm to 0.25 mm) improves the physical properties of a growing media but it may result in reduced drainage and aeration. Sand is the least expensive among all inorganic amendments, but it is heaviest which may result in prohibitive transportation cost.

Bark Rockwool

Polystyrene foam Sand

Polystyrene beads Polystyrene flakes Rice Hulls

Crushed Rice Hulls Calcined clays

J. **Rice hulls:** Rice hulls are the by-product of rice milling industry. It is extremely light in weight and resistance to decomposition. It can also be used effectively in improving drainage of the growing media. However, N depletion is not much serious problem in media amended with rice hulls.

K. **Calcined clays:** Calcined clays are pottery like particles formed by heating montmorrillonitic clay minerals to approximately 690ºC. They are 6 times as heavy as perlite, having relatively high cation exchange as well as water holding capacity. It is very durable and useful amendment. This inorganic soil amendment increases the number of large pores which reduces water holding capacity and improves drainage and aeration.

L. **Bagasse:** Bagasse is a waste by-product of sugar industry. It may be composted or shredded to produce material which can increase the aeration and drainage properties of media. Bagasse contains high sugar which results in rapid microbial activity after the incorporation into media. This decreases the durability and longevity of bagasse and influences nitrogen levels. Although, bagasse is readily available at low cost (usually transportation), but its use is limited.

M. **Coco-peat:** It is a by-product of coconut husk which is prepared by grinding the husk. Presently, it is commercially being used in the pot cultivation of vegetables in the polyhouse as well as for nursery production. It is available in loose form as well as in compressed bricks. It swells 15-18 times more than that of its original weight when soaked in water. It is cheaper and available in abundant quality. Coco-peat is considered best in providing aeration, drainage and life to media.

Bagasse

Cocopeat

Wood shavings

Sawdust

N. **Sawdust and shavings:** Sawdust and shavings are the by-products of lumber mills. These materials get decomposed at a faster rate than bark. It has wider C: N ratio (1000:1). Saw dust is composted with additional nitrogen for one month to increase the quantity of nitrogen. It continues to decompose during the use in pots or greenhouse benches. It is close to neutral in pH when thoroughly composted.

Advantages of Soil-less Media

1. Growers who do not have their own field are also able to grow vegetables and other crops.
2. Soil-less media is free from pathogens and nematodes.
3. It has wide adaptability to produce vegetables in greenhouses, kitchen gardens, in the houses and in the areas where suitable soil is not present.
4. It provides good aeration and drainage.
5. Nutrient control: complete analysis of soil-less culture media is possible, which economizes the fertilizer use.
6. Labour economics: Less handling time, sterilization, weeding, fertilizer application and watering.
7. Reduced transportation cost due to light weight.

19

SOIL PASTEURIZATION

Soil-based media under greenhouse may contain harmful disease causing organisms, nematodes, insect-pests and weed seeds. Hence, soil pasteurization or sterilization for control of soil-borne pathogens should be an essential component in soil-based media under greenhouse. Methods of disinfection like solarization, steam, chemical and biological control can effectively be used.

Soil pasteurization, fumigation, and solarization can be used to control soil-borne diseases, insects, nematodes and weeds. It is important to take steps to prevent pests from being re-introduced to treated soil. Steam is the most common form of heat used for soil pasteurization. Soil fumigation involves the use of volatile chemicals that produce a toxic gas when incorporated into the soil. Fumigants are general biocides; they are effective against fungi, bacteria, nematodes, soil insects and weed seeds. Soil solarization is an environmentally friendly method of using solar heat to control the pests such as bacteria, insects and weeds in the soil. By placing transparent plastic sheets over moist soil during periods of high ambient temperature, the radiant energy can be absorbed and trapped by the soil, thereby heating the top soil layer. The sun heats the soil to temperatures that kill bacteria, fungi, insects, nematodes, mites and weeds. Control of soil-borne pests will improve plant appearance, quality and vigor, crop yields and ultimately profitability.

Solarization

The soil or growing medium can be disinfected during warmer months of the year by covering soil with transparent plastic. This will increase the heat of the soil to a great extent and destroy many soil-borne pathogens and insects. It is based on trapping solar irradiation by tightly covering the wet soil, usually with transparent polyethylene or other plastic sheets. This results in a significant elevation (10-15°C above normal, depending on the soil depth) of soil temperatures up to the point where most of the pathogens are vulnerable to heat when applied for 4-6 weeks and controlled either directly by the heat or by chemical and biological processes generated in the heated soil.

Solarization is a form of soil pasteurization whereby solar energy is trapped beneath plastic sheets spread on the soil surface. It is a cheap and effective means in controlling many important soil-borne fungal and bacterial plant pathogens including those causing *Verticillium* wilt, *Fusarium* wilt, *Phytophthora* root rot, damping-off, crown gall disease, tomato canker and many others. A few heat tolerant fungi and bacteria are more difficult to control with solarization. Soil solarization can be used to control many species of nematodes. Soil solarization controls many of the annual and perennial weeds. A smooth bed, with clods and litter raked away, is best because the plastic will lie snugly against the soil and have fewer air pockets. Air pockets between the plastic and the soil can greatly reduce soil heating and promote "sailing" of the plastic in the wind.

Soil Solarization

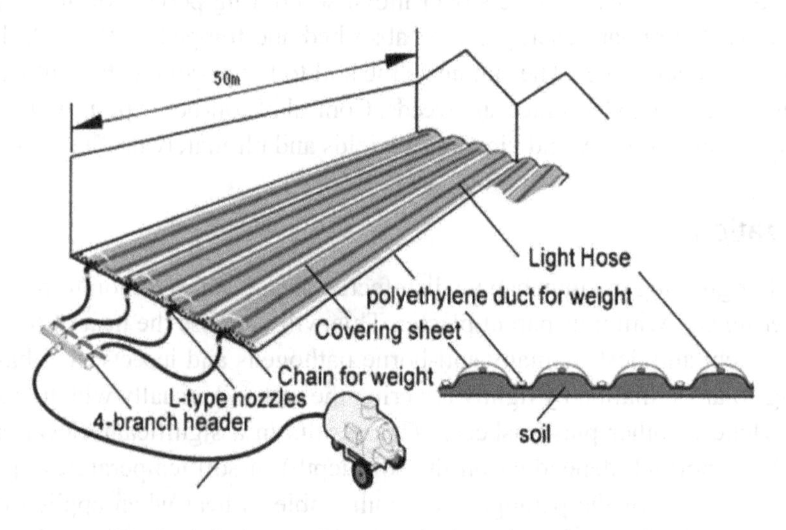

Typical diagram of steam pasteurization system

Steam Pasteurization

Water steam for soil fumigation has been utilized since the beginning of the 20th century, records of which were observed in Germany during 1888. Since then, a wide variety of steam machines have been built to sterilize greenhouse soils. The effect of water steam as a phytoiatric mean is obtained by heating the soil to the levels that cause protein coagulation or enzyme inactivation. The time of exposure to the application and the temperature build up in the soil/substrate are the key parameters for achieving effective results. Such parameters are affected by several factors linked with the soil characteristics (*e.g.* texture, content of organic matter, water content *etc.*) and technical aspects of the application (*e.g.* steam distribution system). Although, high temperature steaming at 85°C to 100 °C in early days killed too many beneficial soil organisms (*i.e.* mycorrhizal fungi, nitrogen fixing bacteria) along with the pathogens and led to the production of phytotoxic compounds harmful to crop plants. Therefore, currently an effective treatment shall achieve a soil temperature of at least 70 °C for 30 minutes. A less drastic method has also been proposed with a temperature of 50-60°C for few minutes. Therefore, to make this technology effective growing medium should first be loosened before pasteurization. This will help the movement of steam though the pores and transmit the heat rapidly within the medium. The root medium should not be dry and if dry, addition of water speeds up the rate of pasteurization. The excess watering may slow down the speed of pasteurization, therefore moistening of root medium a week or two prior to pasteurization is the best procedure, which breaks the dormancy of many unmanageable weed seeds and then pasteurization destroys them easily. **To strengthen the adoption of this technology, Govt. of India is also providing 50% subsidy under a Mission for Integrated Development of Horticulture (MIDH).**

Steam pasteurization

Fumigation or Chemical Pasteurization

Soil fumigation is a chemical control strategy used independently or in conjunction with cultural and physical control methods to reduce populations of soil organisms. Soil fumigants can effectively control soil-borne organisms such as nematodes, fungi, bacteria, insects, weed seeds and weeds. Different fumigants have varying effects on the control of these pests. Some are pest-specific, while others are broad spectrum biocides and kill most of the soil organisms. Because of treatment costs, applicators use soil fumigants primarily on high valued crops such as vegetables, ornamentals and fruits. Very careful attention must be paid to the details of how the fumigation is done in order to guarantee excellent control of the targeted organisms and complete safety for workers and other persons close to the treated area, while also limiting the impact the product could have on our environment.

Many factors affect soil fumigation and its effectiveness for pest control. The pest and its habits will affect fumigant selection, application rate, fumigant placement and necessary length of exposure. Soil factors also play a key role in fumigation. Soil texture, soil condition, debris, soil moisture and soil temperature may affect the volatility, movement and availability of the fumigant once applied. Fumigant dosage is both pest and soil-dependent. After fumigation, aeration is important to make sure phytotoxicity does not occur.

The procedure adopted for soil-pasteurization by each chemical is tabulated below:

Chemical	**Dose**	**Recommendation**
Formaldehyde	75 L /1000 m^2	Irrigate the soil to the field capacity. Drenching soil with Formaldehyde (37-41%) at 1:10- Chemical: water ratio. Covering soil with black polyethylene sheet and keeping it for a week. Removal of covering after 4-5 days and irrigation of soil to drain excess of fumes, if any.
Metham sodium	30 L/1000m^2	Cover the soil after application of chemical for 4-5 days with black polyethylene sheet and aerate for 8-10 days.
Dazomet	40 kg/1000m^2	Pre-plant application (30 days before sowing/ planting.)
Chloropicrin (Tear Gas)	30-45 L/1000 m^2	Cover soil for 1-3 days with gas proof cover after sprinkling with water. Aerate for 14 days or until no odour is detected before using.
Basamid	70-75 kg/1000 m^2	Cover soil for 7 days with gas proof cover and aerate for a week before use.

Physical propagation facilities such as the propagation room, containers, flats, knives, working surface, benches *etc.* can be disinfected using 1 part of formalin in 50 parts of water or 1 part sodium hypochlorite in 9 parts of water. Care should be taken to disinfect the seed or the planting materials before they are moved into the greenhouse with a recommended seed treatment chemical for seeds and a fungicide-insecticide combination for cuttings and plugs. Disinfectant solution such as trisodium phosphate or potassium permanganate placed at the entry of the greenhouse would help to get rid of the pathogens from the personnel entering the greenhouses.

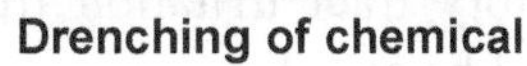

Drenching of chemical | **Covering soil with plastic for 4-7 days** | **Removing excess fumes from soil via irrigation**

Chemical sterilization of soil (First method of chemical application)

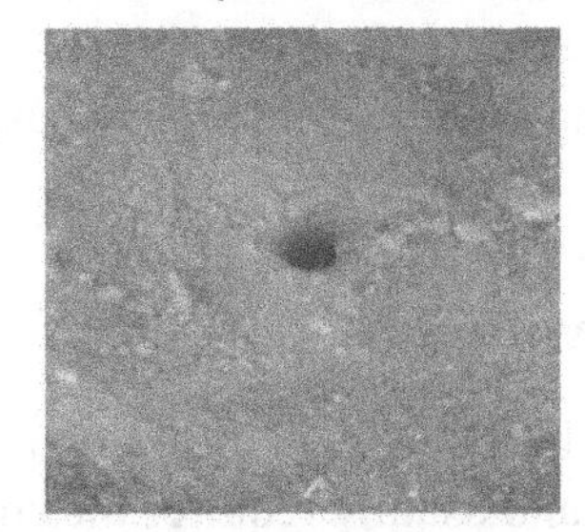

Prepartion of holes in soil media (Second method of chemical application)

Chemical sterilization in soil media

Safety Directions While using Chemical Sterilants

- Harmful if inhaled or swallowed. Will damage the eyes, nose, throat and skin. Repeated exposure may cause allergic disorders. Interacts with alcohol, so avoid alcohol on day (s) of use. Do not inhale vapours or spray mists. The fumes first cause smarting and then watering of eyes. This should be taken as a warning sign. The liquid can cause burns. Use and store these chemicals in well ventilated areas.
- While opening the chemical container and preparing product for use, wear chemical resistant clothing buttoned to the neck and wrist, a hat, elbow-length chemical resistant gloves, chemical resistant footwear, a full face-piece respirator with organic vapours.
- Workers previously experiencing skin or respiratory tract irritation from chemical (s), their exposure to such products should be avoided.
- After mixing with water, do not allow mixture to stand as poisonous fumes are released on standing.
- Workers manually sealing should wear the personal protective clothing specified for applicators.

Augmentation of Soil with Bio-agents

Augmentation refers to all forms of biological control in which natural enemies are periodically introduced and usually requires the commercial production of the released agents. Bed preparation by building up rich flora of biological control agents like *Trichoderma* spp., *Pseudomonas fluorescens, Paecilomyces lilacinus etc.* is essential for the management of soil-borne pathogens especially nematodes. Additionally one should adopt the following precautions:

- Avoid/ repair faulty greenhouse structures which help in the entry of insect-pests.
- Always use insect-proof net screens.
- Greenhouses along with workers/visitors should have double entry gates preferably in L shape so as to minimize the risk of pest entry.

Augmentation of FYM can be achieved by mixing *Trichoderma harzianum* culture @ 1 kg/500 kg FYM and/or *Pseudomonas fluorescens or Paecilomyces lilacinus* @ 50 g/m^2 15 days prior to sowing.

20

TESTING SOIL AND WATER SUITABILITY FOR GREENHOUSE CROPS

Soil Fertility

Soil fertility is a complex characteristic governing the availability of nutrients in soil buffer, which can also be managed by the external application of nutrients organically or inorganically. The capacity of soil to provide nutrients to plants instantly is considered for short term productivity, while various physical and chemical properties of soil are considered substantially dealing soil fertility in terms of the highest level of productivity. To achieve sustainable production in greenhouse system, all the interactions among physical, chemical and biological characteristics must be taken into account, as these parameters directly or indirectly influence the dynamics of nutrients and their availability in the plant system

Therefore, it become very important to optimize plant nutrition under protected culture, when a grower is practicing crop cultivation on soil-based media after reviewing the short, medium or long term objectives for greenhouse production system. Hence, it is significantly important to analyze the presence or absence of many factors contributing towards physical, chemical and biological properties of soil. For instance, Mycorrhizal fungi, *Rhizobium*, Azotobactor etc. are the important beneficial factors determining biological activities of the soil, while soil-borne pathogens like nematodes, Fusarium, Ralstonia, Rhizoctonia etc. are the harmful factors deciding the fate of crops cultivated in such soils. Soil physical properties like texture and structure must be given due importance for crop cultivation in soil media under protected environment. Similarly, water holding capacity gives an idea for proper water management. The chemical factors of soil like nutrients status, organic matter, soil pH and soluble salts plays vital role to strategize nutrient management programme

Soil Texture

Texture indicates the relative content of particles of various sizes, which include sand (0.05–2 mm), silt (0.002–0.05 mm) and clay particles (< 0.002 mm) in soil **(Fig. 20.1)**. Texture influences the ease with which soil can be worked, the amount of water and air it holds, and the rate at which water can enter and move through soil.

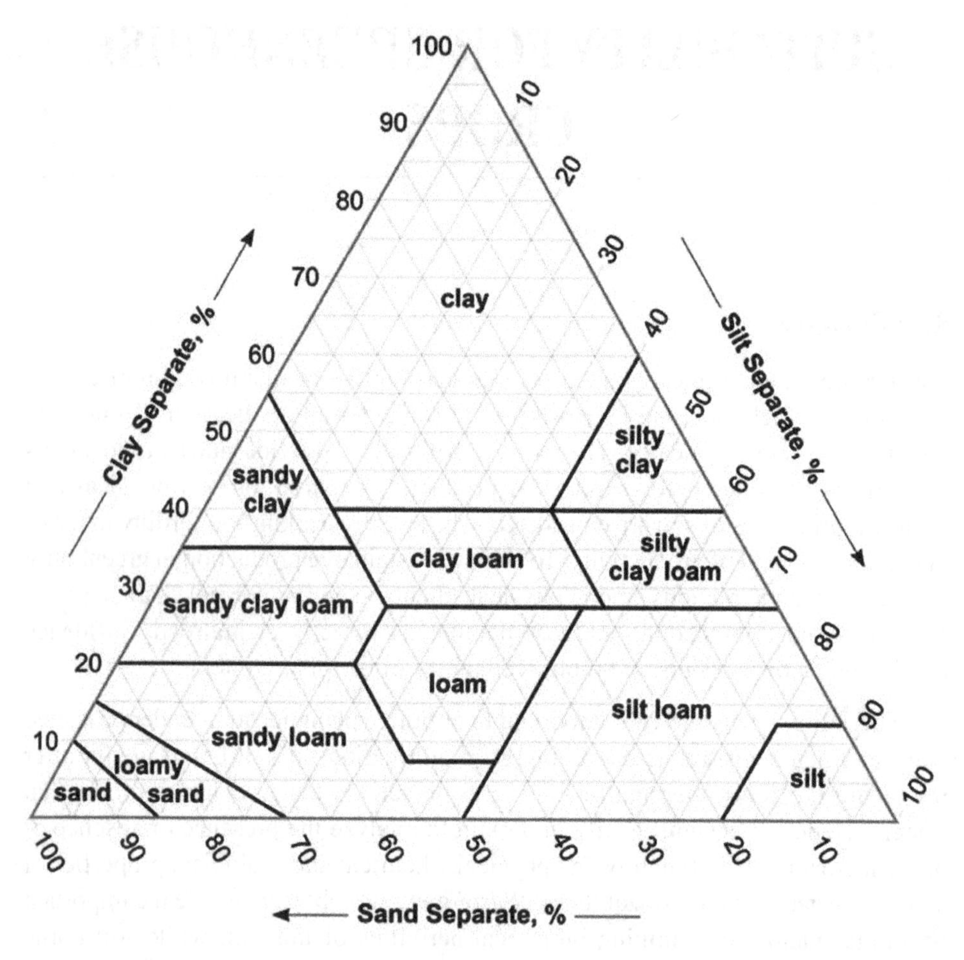

1. Sand 2. Loamy sand 3. Silt 4. Sandy loam 5. Loam 6. Silt loam 7. Sandy clay loam 8. Clay loam 9. Silty clay loam 10. Sandy clay 11. Silty clay 12. Clay

Fig. 20.1: USDA Texture Triangle

Sand, loamy sand, silt, sandy loam, loam, silt loam, sandy clay loam, clay loam, silty clay loam, sandy clay, silty clay and clay 12 different classes of soil described based on the texture.

Soil structure

Soil structure describes about the aggregation of particles of sand, silt, and arrangement of maco and micro-pores within soil particles. Soil structure has direct influence on air-water movement and their availability and, biological activities, which ultimately decide the overall performance of a crop.

Soil Organic Matter (SOM)

Soil organic matter (SOM) consists of 3 components *viz.*, livening biomass of micro-organisms, well-decomposed organic matter and highly stable organic material. The crop residues added into the surface of soil are not considered as part of SOM.

On the addition of organic material in the soil, the degradation process gets started with the involvement of micro-organisms and completed within a period of few weeks for protein, while lignin content of organic material takes months together for the completion of degradation process.

SOM improves the release of plant nutrients, regulate energy supply for soil micro-organisms and brings improvement porosity and soil aeration, increase of water-holding capacity in sandy soils and limit the compaction and erosion of heavy soils, thereby improving soil structure. SOM is estimated by the following equation:

$$\text{Organic matter (\%)} = \text{Organic carbon (\%)} \times 1.724$$

Based on the organic matter content, soils are rated as very low, low, medium and high as depicted in **Table 20.1.**

Table 20.1: Soil rating based on organic matter content

Rating	SOM (%)		
	Sandy Soil Class (1, 2, 4)	Loamy Soil Class (5, 6, 7, 8)	Clay and Silty Soil Class (3, 9, 10, 11, 12)
Very low	< 0.8	< 1.0	< 1.2
Low	0.8–1.4	1.0–1.8	1.2–2.2
Medium	1.5–2.0	1.9–2.5	2.3–3.0
High	> 2.0	> 2.5	> 3.0

Soil pH

pH of soil is also one of the pre-requisites for successful cultivation of crops under protected environment. It is a numerical measure of the acidity or alkalinity in soils, usually measured on a scale from 0 to 14. Neutral soils have a pH of 7, soils with pH value of 0are categorized as highly acidic, while pH of 14 describes about highest level of alkalinity in soil. Most of the crops thrive well in the pH range of 6.0 to 7.5 and it is well established fact that pH affect the microbial activity. The pH values more than 7.5 are considered regarded as undesirable as it can lead to precipitation of Ca and Mg carbonates as well as orthophosphates, causing clogging in micro-irrigation system (MIS). So, the availability of nutrients to growing plants is highly correlated with pH of soil as well as of nutrient solution. The relation between pH of media and nutrient availability is depicted in **Fig. 20.2.**

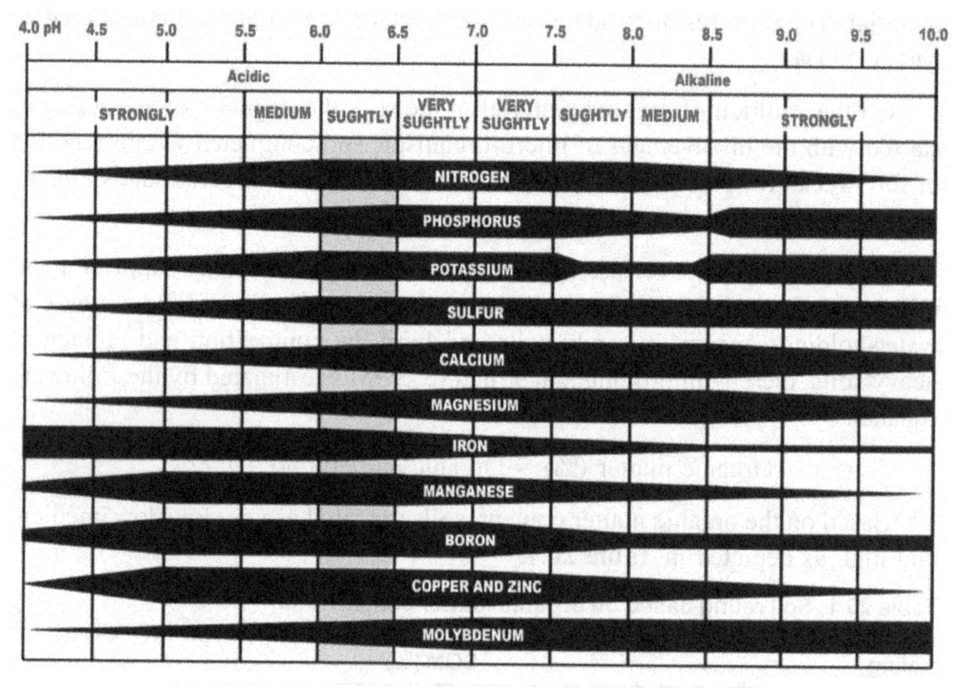

Fig. 20.2: Relation between pH of media and nutrient availability

Rapid adjustment of pH

The corrective measures to adjust pH of rooting media rapidly in pots or benches containing plants.

To Lower pH

- Apply iron sulphate @ 450 g to 1125 g per 400 litres of water to the rooting media.

To Raise pH:

- **Option 1:** Add 450 g of hydrated lime in 400 litres of water and allow it to settle down. The clear solution can be applied to the rooting medium.
- **Option 2:** Firstly, mix 450 g of hydrated lime in 12 to 20 litres of warm water in a plastic bucket. Allow the mixture to settle and pour off the clear solution into another plastic bucket, and then apply the clear solution through MIS.

Soluble Salts

The presence of high salt concentration in growing media affects various physiological processes of plants and the deposition of salts in laterals of MIS causes clogging of the system. Electrical conductivity (EC) is a measure of the levels of soluble salts present in the media, which is related to the quantity of dissolved ions (mostly from fertilizer salts) in solution. Excessive soluble salt levels in the growing media are most frequently caused by over application of nutrient salts i.e. fertilizers; however levels could also become elevated due to poor drainage, inadequate watering and leaching, or when the functioning of root system is impaired. The higher level of EC may cause root injury, leaf chlorosis, burning of leaf margins and/or sometimes wilting of plants.

Units for the measurement of EC and relation between them are given below:

1 millimho per centimeter (mmho/cm) = 1 deciSiemen per meter (dS/m) = 1,000 micromhos per centimeter (µmhos/cm) = 0.1 Siemen per meter (S/m)

TDS (ppm) = EC (dS/m) x 640, for EC between 0.1 and 5.0 dS/m	TDS (ppm) = EC (dS/m) x 800, for EC > 5.0 dS/m
690ppm / 640 = 1.08 dS/m (EC) 1.08 dS/m x 640 = 690 ppm or TDS	690ppm / 800 = 0.86 dS/m (EC) 0.86 dS/m x 800 = 690 ppm or TDS

Note: 640 for EC <5.0 dS/m and 800 and 800 for EC>5.0 dS/m are for EC>5.0 dS/m are just conversion factors based on total salt content

Soil Testing

A soil test is important for several reasons:

- To optimize crop production
- To protect the environment from contamination by runoff and leaching of excess fertilizers
- To aid in the diagnosis of plant culture problems

- To improve the nutritional balance of the growing media
- To save money and conserve energy by applying only the amount of fertilizer needed.

Pre-plant media analyses provide an indication of potential nutrient deficiencies, pH imbalance or excess soluble salts. This is particularly important for growers who mix their own media. Media testing during the growing season is an important tool for managing crop nutrition and soluble salts levels. To use this tool effectively, you must know how to take a media sample to send for analysis or for in-house testing, and be able to interpret media test results.

Important Properties of Soil Media

- There should be a good balance between water holding capacity and porosity of soil.
- The drainage capacity of media should be good (poor drainage and aeration in highly compact media results in poor root growth, while low water and nutrient holding capacity of highly porous media affects growth and development of plants).
- The optimum range of pH is 5.0 to 7.0 (occurrence of toxicity of micronutrients such as Fe, Zn, Mn and Cu, and deficiency of major and secondary nutrients in media having pH <5.0, while a high pH (>7.5) causes deficiency of micronutrients including boron.
- The desirable range of EC of media is 0.4 to 1.4 dS/m.

Corrective Measures for pH Adjustment

- Soil amendments like lime (calcium carbonate) and dolomite (Ca-Mg carbonate) and basic fertilizers like calcium nitrate, calcium cyanamide, sodium nitrate and potassium nitrate can used to raise the pH to a desired level.
- Amendments like sulphur, gypsum and Epsom salts and acidic fertilizers like urea, ammonium sulphate, ammonium nitrate, mono ammonium phosphate and aqua ammonia and acids like phosphoric and Sulfuric acid can be used to reduce high pH of the media.

Checklist

1. Conduct pre-plant media analysis to provide an indication of potential nutrient deficiencies, pH imbalance or excess soluble salts. This is particularly important for growers who mix their own media.

2. Conduct media tests during the growing season to manage crop nutrition and soluble salts levels.
3. Always use the interpretative data for the specific soil testing method used to avoid incorrect interpretation of the results.
4. Take the soil sample for testing about 2 hours after fertilizing or on the same day. If slow release fertilizer pellets are present, carefully pick them out of the sample.
5. In a greenhouse where a variety of crops are grown, take soil samples from crops of different species.
6. If a problem is being diagnosed, take a sample from both normal and abnormal plants for comparison.
7. Be consistent in all sampling procedures each time you sample.

Soil Sampling (Fig. 20.3):

It is always recommended to draw soil samples prior to the initiation of crop production programme and get relevant information on nutrient status of soil to execute fertigation appropriately in greenhouse crops. Such an initiative will definitely help to use fertilizers judiciously as well as avoid any kind of nutrient deficiency in a crop under cultivation.

The soil sample must represent the soil media being used for crop cultivation. The following steps are very important to draw a representative sample for improving the accuracy of soil test:

- Samples can be taken with a soil probe, pipe, auger or hand trowel.
- Remove the top debris, residue or turf thatch from the soil.
- Sample soil sample upto the depth of 15-20 cm as illustrated in Fig. 20.3.
- Randomly zigzag over the area collecting 10- 15 cores or slices (primary samples) in a clean plastic bucket for an area of 1000 square meter.
- Each core or slice should be taken at the same depth and volume at each site.
- Break up lumps and thoroughly mix all the primary samples to constitute a composite sample.
- Divide the composite sample into 4 equal parts to draw true representative of soil.
- Discard 3 parts and take representative sample for analysis of soil parameters (Soil EC and pH).

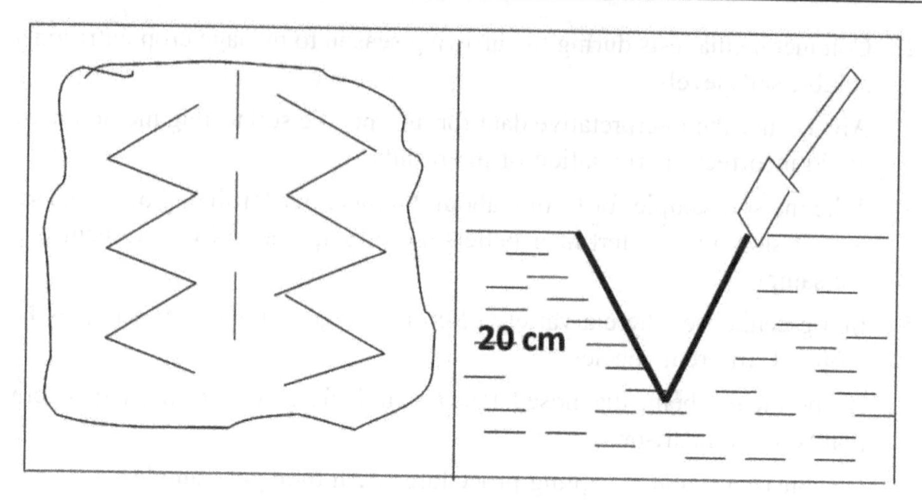

Fig. 20.3: Methodology for Soil Sampling

Methodology to Assess EC and pH of Media

Saturated Media Extract Method

The Saturated Media Extract Method (SME) was developed at Michigan State University and has been routinely used in their soil testing lab. It allows extraction of moist samples just as they come from greenhouses. Drying of samples is unnecessary and undesirable. Storage of prepared root media in either the dry or moist state will influence the soluble nitrate-nitrogen (N) and soluble salt levels. If samples will not be extracted within two hours of receipt, store them under refrigeration (~4°C).

Soluble salt guidelines for greenhouse growth media using various media to water ratios.

1:2 or 1: 5 dilution method

In this test an air dried sample of soil and water are mixed together at 1:2 or 1:5 ratio on volume basis. The liquid extract is then separated from the solids using laboratory grade filter paper or a common coffee filter. The extract is the ready for analysis. This is very easy to master and quite suitable for on-site greenhouse testing of pH and soluble salts using hand held pH and EC meters or pens.

The reading of 1:2 or 1: 5 dilution methods can be interpreted as per the details given in Table 20.2.

Table 20.2: General Interpretation guidelines for greenhouse growth media analyzed by the Saturated Media Extract Method

Soluble Salts (mS/m)			
Saturated Media Extract Method	**1 part media to 2 parts water**	**1 part media to 5 parts water**	**Interpretation**
0.00 to 0.74	0.00 to 0.25	0.00 to 0.12	Very low salt levels. Indicates very low nutrient status.
0.75 to 1.99	0.25 to 0.75	0.12 to 0.35	Suitable range for seedlings and salt sensitive plants.
2.00 to 3.49	0.75 to 1.25	0.35 to 0.65	Desirable range for most established plants. Upper range may reduce growth of some sensitive plants.
3.50 to 5.00	1.25 to 1.75	0.65 to 0.90	Slightly higher than desirable. Loss of vigor in upper range. OK for high-nutrient requiring plants.
5.00 to 6.00	1.75 to 2.25	0.90 to 1.10	Reduced growth and vigor. Wilting and marginal leaf burn.
6.00+	2.25+	1.10 +	Severe salt injury symptoms with crop failure likely.

This method can also used for on-site estimation of pH of the media, interpretation of which can be traced from Table 20.2.

General Rules for Using the pH Meter

1. Read carefully the instructions manual provided with the instrument.
2. The reading bulb (electrode) must remain constantly moist. It should therefore be stored in water (not distilled water) or in specific storage solution (it may be sufficient to place a moist ball of cotton wool in the bulb protection cover).
3. The calibration should be checked frequently by immersing the electrode in the specific known pH buffer solutions (generally pH 7.0 and 4.0). If the reading differs noticeably from the nominal value (an error of 0.1–0.2 is tolerable for field measurements), it needs to be calibrated again following the instructions in the manual.
4. If it takes a long time to get a stable reading, it is recommended to clean the electrode thoroughly using paper soaked in water and washing it in plenty of water (specific wash solutions are also available on the market). If the readings are still unsatisfactory after doing this, it may need replacing.
5. Store the instrument in a warm, dry place.

General Rules for Using the EC Meter

1. Read carefully the instructions manual provided with the instrument.
2. The EC value* depends heavily on the water temperature, so much so that when expressing the results it is important to indicate the reference temperature (usually 25 °C). Most of the instruments on the market – including the relatively affordable models – come with an automatic temperature compensation device. This means that the EC and the temperature values are measured and the reading at the reference temperature is provided automatically. If the instrument has this device, the readings can be used without further calculations. If it does not, the temperature needs to be taken manually and the reading needs to be converted during the calibration phase.
3. The calibration should be checked frequently by immersing the electrode in the specific standard solutions (available at different concentrations). If the reading differs noticeably from the nominal value (an error of 0.1–0.2 mS/cm is tolerable for field measurements), it needs to be calibrated again following the instructions in the manual.
4. The electrode must be cleaned periodically.
5. Store the instrument in a warm, dry place.

* Electric conductivity measurements do not indicate the relative amounts of any specific salt and ion. Additional specific tests typically run by designated laboratories must determine concentrations of specific ions.

21

PLUG TRAY: A COMPONENT OF HIGH TECH NURSERY

Plug Tray Nursery

Plug trays are now-a-days commonly used for healthy nursery raising, which are popular among nurserymen with the names of plastic trays or pro trays. These trays usually have different sizes of cells/plugs and accordingly different crops species respond variably to cell sizes of plug trays. Looking to the cost of seeds, nursery raising in plug trays ensure better germination, care, survival and ultimately make it certain to provide uniform and healthy growth to seedlings in comparison to the seedlings raised through traditional methods of nursery raising on flat/raised raised bed. Plug tray nursery raising also provides an opportunity for better handling and storing at farm and during transportation and is more reliable and economical.

Rooting Media

Soil has traditionally been accepted as part of seedling growing mixture. The amount of undesirable material coming with the soil such as diseases, weeds, trash and clay is high. Soils may contain unknown plant toxins, extremes in soil pH variations, nutrient content or unknown chemical residues. The alternatives to soil, either in part or in total are not without problems. Mostly artificial soil-less media is used for raising healthy and vigorous seedlings in plastic protrays.

There are different types of soilless media available which can be used for plug tray nursery raising. Some soil-less materials used as rooting media are given below:

Coco peat: It is completely free from infestation of any pest or pathogen. It is commonly used as a medium for raising nurseries of vegetables and ornamental plants.

Perlite: It is light rock material of volcanic origin. It is essentially heat expended aluminum silicate rock. It improves aeration and drainage. Perlite is neutral in reaction and provides almost no nutrients to the media. As an inert light material, perlite is used to reduce bulk density. It does not retain moisture or hold any plant nutrient. There is no real value of having perlite as part of the growing media.

Vermiculite: It is heat expended mica. It is very light in weight and has minerals for enriching the mixture (magnesium and potassium) as well as good water holding capacity. Neutral in reaction (pH), it is available in grades according to sizes. It is able to reduce the loss of nutrients through leaching.

Cocopeat Perlite Vermiculite

Pine bark: Two key disadvantages with pine bark are the effect of its toxins on plant growth and in its hydrophobic (or difficult to wet) nature. There have also been problems at times with purity of sample and in uniformity of particle size following milling. For seedling mixes, bark should be milled to a maximum particle size of 5 mm. Most of the toxin content can be leached by ageing in open air stacks, however the bark must be wet first. Newly stripped bark is harder to wet following drying than stockpiled and weathered bark. Pine bark can be composted with the addition of small amounts of slow release fertilizers to help in the process. This process can give some control over certain diseases which may contaminate the bark. In open stacks, the heat generated throughout the heap will not be even and therefore in the cooler areas, the temperature will not rise to the thermal death point of many diseases.

Hardwood sawdust: There are similar problems with sawdust to those of pine bark *i.e.* some timbers do have quite high toxin levels. Sawdust composted in the recommended manner as seen in bark treatment, can be combined with pine bark and with peat moss. It is desirable to keep the percentage by volume of sawdust to around 40 % or less.

Softwood sawdust: The most common softwood sawdust available is from *Pinus radiata*, which contains less toxins than some of the Australian hardwoods.

Pine bark

Hardwood sawdust

Softwood sawdust

Sand: The value of adding sand as an inert mineral to any seedling mix is to obtain a balance between the other components and to give some bulk to the final product. A common belief is that sand helps with drainage in fact the reverse can occur. For example, if added to peat moss it may fill in many of the spaces between peat particles, thereby reducing loss of moisture. Also re-wetting of a dried out mix is improved if there is a small amount of sand present. However, over mixing of peat and sand in a rotary mixer can lead to pulverizing of the peat, thus changing its structural qualities. On the other hand large amounts of sand added to a seedling mix is likely to produce rather dense mixtures. Likewise, there needs to be a balance between the extremes of fine and coarse sands.

Polystyrene pellets or beads: This material is used sometimes, if for no other reason than to add bulk. It has no nutrient value or water holding capacity. Polystyrene is difficult to handle outside when there is a wind. Although the results are better if mixed dry with a little brown coal prior to water addition.

Sand

Polystyrene pellets or beads

Brown coal: This material is known also as Lignum Peat or coal fines. The ratio of brown coal in a soil-less mix to other components needs to be low, about 20% by volume or even less. Higher volumes with adequate nutrition may cause plants to respond with lush growth and will be difficult to harden. Brown coal has a high water holding capacity with a high level of unavailable water. Its water loss is in fact high too and it has been shown that plants transpire less water growing in brown coal than is the case with peat moss.

Peat moss: The use of peat moss for horticulture is increasing worldwide. In Australia, there are a number of peat deposits, but not many are suited for seedling mixtures. The fine sedge peats are generally not suited to seedling mixtures due to their dense nature. Most peats are relatively stable against rapid decay. They are valued for their high water holding capacity. The sphagnum peats have a better balance between water and air than most sedge peats. Even though peats have this capacity to hold water, but loss of moisture is high through transpiration and evaporation. Because of this, it is claimed that plants growing under optimum conditions would grow faster in this material.

Brown coal

Peat moss

Making a Suitable Nursery Media

Mainly three ingredients namely cocopeat, vermiculite and perlite are used as rooting mix/medium for raising nursery. These ingredients are mixed in 3:1:1 (Volume basis) ratio before filling in plastic plug trays. It has also been critically noticed that healthy planting materials can be raised in coco peat alone, which is as good as raised in three constituents of rooting media. Coco peat from Sri Lanka and India contain several macro and micro-nutrients, including substantial quantities of potassium, sodium and chloride. So, before using cocopeat as rooting media, it is always very important to wash cocopeat with good quality water 2-3 times to remove excess of elements which are soluble in water such as potassium, sodium and chloride. Washing of cocopeat helps to reduce the electrical conductivity to a tolerance limit. Thereafter, buffering of coco peat is done with calcium nitrate @ 100g per 10 litres of water for 5 kg of cocopeat. During this process, calcium [2^+] is introduced in order to remove monovalent positive ions such as potassium [1^+] from the coconut complex. In this way, we can remove not only elements which are soluble in water but also elements which are bound to the coconut complex. Ideally, the treated water should be administered into coco-peat over a 24 hours period via a slow splinker system if possible, otherwise cocopeat can be rested in calcium nitrate solution for 24 hours. Once the resting period is over, coco-peat is rinsed with water twice and then cocopeat becomes ready to use as rooting medium for nursery production. Now-a-days, buffered cocopeat is also available in the market, which can be used directly without following the above mentioned process of washing and buffering.

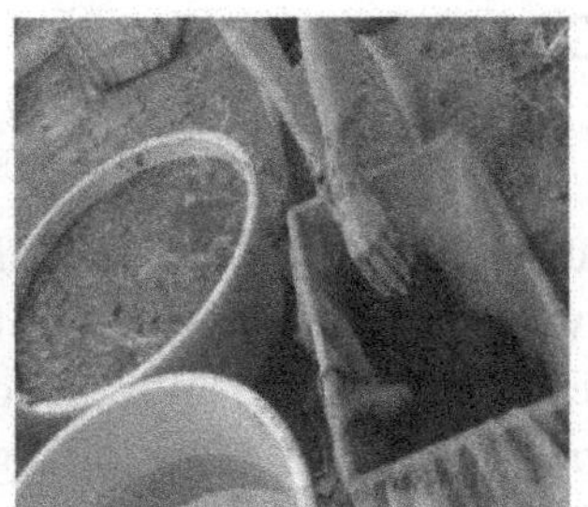

Removal of excess salt through washing and buffering of cocopeat with $CaCO_3$

Selection of a Seedling Tray

Plastic trays of different sizes are available in the market for nursery production, which are also popular known by the names of plug trays or pro-trays.

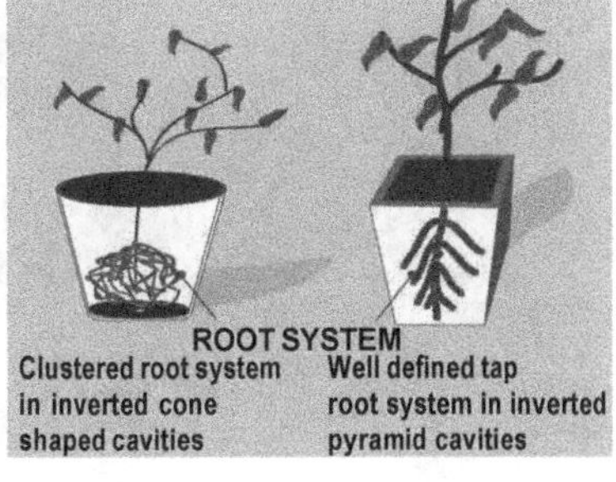

Plug tray with inverted coneshaped cavities **Plug tray with inverted pyramidshaped cavities** **Comparison of root development in inverted cone and pyramid shaped cavities**

The selection of the seedling tray depends on size of seeds, trays with 102 or 104 cavities (or plugs) having less cell volume are mostly preferred for small seeded crops like tomato, chilli, brinjal *etc.*, whereas trays with 70 or 50 cavities having more cell volume are used for large seeded crops like most of the cucurbits, papaya. Based on the size of plug cavities, plug trays with inverted cone and inverted pyramid type cavities are available in the market. It has been established that plug trays with inverted pyramid shaped cavities are very much suitable for ideal and strong growth of root and shoot system because these types of cavities offer a wider space for the development of root system. In contrary, inverted cone type cavities have limited space for proper growth and extension of roots thus cause clustering of root system.

METHODS OF SOWING

Manual Sowing

Most of the nurserymen usually follow the practice of seed sowing with hands. However, sowing of seeds with the help of a dibbler helps to maintain uniform sowing depth and speed as seed dibbler make uniform holes in the rooting medium, which is otherwise impossible while attempting sowing with hands. The dibbling depth is recommended as 3 times the seed diameter. Once the seed sowing is over, a thin layer (1-1.5 times the diameter of the seed) of either vermiculite or coco peat is spread over the surface of sown seeds. Manual sowing is a slow process and one can fill approximately 20-25 trays in an hour, hence not recommended for large scale nursery production.

Manual sowing

Machine Sowing

A seeder assembly can be used for automated tray filling, sowing, covering and watering for large scale nursery production, which make it possible to fill about 90-100 trays in an hour.

AFTER CARE

Plug Tray Placement

It is important to transfer the sown plug trays to a darkened germination room/ closed chamber maintained at a mild temperature. Small scale nurseries may not have space for a separate germination room, so one can use one of the corners for this purpose. This step is very much needed to retain sufficient moisture for germination by reducing evaporation losses. The practice of stacking many plug trays on top of each other is also followed by some nurserymen but this practice is not recommended as plug tray at the top will compress the trays beneath. Due to space constrains, germinating plug trays need to be stacked sometimes. So, it is recommended to use a ziz-zag arrangement of placing trays one over anther, which will enable trays to avoid direct contact with the media of lower one. Commercial nurseries follow the practice of putting germinating trays on nursery tables, which can be managed very efficiently by few persons.

Use of machines for sowing

"Indigenously developed sowing machines (IIHR, Bengaluru)"

Dibbler cum seeder for plug trays

Automatic plug tray filling, dibbling, seeding and watering machine

Watering

Water quality plays very important role in deciding the health and growth of seedlings. The most common problem usually faced by most nurserymen is due to the present of dissolved salts in water. Electrical conductivity (EC) and pH are two important criteria used to adjudge the water quality, the ideal range of which are less than 1 mS cm^{-1} and 6.5-8.4, respectively. Seedling growth gets affected badly, if the water EC is high, while higher pH level is associated with higher levels of salts that will damage seedlings directly. Therefore, assessment of these quality parameters (EC and pH) can easily be done with the help of EC meter and pH meter.

Trouble Shooting Under the Situation of High EC of Water

- Provision for rainwater harvesting should be made and the rainwater can be mixed with available water to reduce the EC to a desirable level.
- Water softeners like potassium chloride can be used to lessen water EC.
- Good drainage in the seedling trays should be ensured as it does not allow build up of salts under the high water EC at initial stage.

Suggested Time and Frequency of Irrigation

It is suggested to irrigate seedlings in the morning hours, because water droplets may remain on seedling leaves which can make seedling prone to fungal infection if irrigated during evening hours. Media used for raising nursery plays very important role in deciding the frequency of watering, if the rooting medium is porous, less water is required. Similarly, weather conditions also decide the frequency of water application, watering can be done 2-3 times a day during summer and it can be restricted to single irrigation during winter season. Likewise, irrigation can be reduced to once a day during damp weather or rainy season. Sometimes, need based irrigation should also be practiced depending upon the health of seedling. For example, when drooping of seedlings is observed, it gives clear indication that seedlings require water hence should be irrigated immediately. During last 4-5 days prior to transplanting, reduce watering to assist in hardening of seedlings.

Nutrient Application

Generally, inert materials are used as rooting media for plug tray nursery raising, so it becomes very important to fulfill the nutritional requirement of growing seedling through nutrient solution. In low-tech or small nurseries, nutrient solution can be applied with the help of water cans fitted with nozzles. In case of hi-tech nurseries, nutrient solution is first prepared in a tank and applied uniformly through boomer via misting. The composition of nutrient solution is largely dependent on the stage of seedlings. Generally, daily application of complex fertilizer (19:19:19) along with micro-nutrients (preferably Grade IV: Fe-4.0%, Mn-1.0%, Zn-6.0%, Cu-0.5%, B-0.5%) @ 5 g per 10 litres of water, starting from the cotyledon stage till 16 days and gradually increasing the nutrition as per the growth stage of seedlings up till around 22 to 24 days. This process of nutrient application can also look after the water requirements of growing seedlings. Thereafter, application is stopped to begin the process of hardening prior to actual transfer of seedlings in the main field. The detailed information on nutrient application in seedlings is indicated in a table below:

Days After Sowing	Composition of nutrient solution (g in 10 litres of water)	
	19:19:19	Micronutrients (Grade IV)
8-16	5	1
16-19	10	2
19-22	15	2
22-24	20	4

Advantages of Plug Tray Nursery Raising

1. Seedling can be raised under adverse climatic conditions, where it is not possible otherwise
2. Healthy seedlings can be raised.
3. There is no chance of soil-borne fungus or virus infection to seedling as the nursery is grown in soil-less media
4. Drastic reduction in mortality in transplanting of seedlings as compared to traditional nursery raising
5. Early / off-season planting is accomplished by raising such nursery
6. Easy transportation
7. Weed free production
8. Better root development
9. Seed rate can be reduced to 30 -40% compared to open field nursery production
10. Saving the fertilizer and water

DO's and DON'Ts in Nursery Management

DO's

- Prefer protected strcuture having provision for double door system.
- Ensure that structures used for nursery production should have been fabricated using nuts and bolts in place of welding.
- Work on a tarpaulin sheet while filling plug trays with media to avoid any kind of direct contact with soil.
- Monitor EC and pH of water as well nutrient solution with the help of EC meter and pH meter.
- While transporting, seedlings trays must be packed in plastic crates to avoid injury to seedlings.

- The plastic crates containing seedlings trays should appropriately arranged in the transport vehicle.
- Spaying of chemicals must be ensured with protective clothing.

DON'Ts

- Do not allow anyone in nursery production area without disinfecting feet.
- Do not stack the freshly sown plug trays tightly and heavily.
- Do not leave gaps in the nursery structure as it can lead to easy entry of insect-pests.
- Do not dump waste near the nursery as it may serve as a breeding place for so many pathogens.
- Do not allow weeds to grow around the nursery area.
- Do not leave any gap while spreading weed mat on the ground as it may lead to weed emergence.
- Do not let water to stagnate in and around the nursery area.
- Do not keep the doors of nursery structure open.
- Do not leave holes in the insect proof net/screen.
- Do not allow the seedlings to come in direct contact with soil.
- Do not fold the seedling trays in the crate while attempting the transport process.
- Do not allow other plants in any case to grow inside the nursery structure.
- Do not overwater the rooting media.

22

BASICS OF FERTIGATION IN GREENHOUSE PRODUCTION SYSTEM

The practice of supplying crops with fertilizers *via* the irrigation water is called fertigation or Fertigation is a process of application of water soluble fertilizers or liquid fertilizers through irrigation water. The nutrients are applied via fertigation directly into the wetted volume of soil immediately below the emitter where root activity is highly concentrated. In fertigation, plants receive small amounts of fertilizer early in the crop season, when plants are in vegetative stage. The dosage is increased as fruit load and nutrients demand grow and then decreased as plants approach to the end of its crop cycle. This gives plants the needed amount of fertilizers throughout the growth cycle rather than a few large doses.

Timing, amounts and concentration of fertilizers applied are easily controlled in fertigation. Fertigation allows the plants to absorb up to 90% of the applied nutrients, while granular or dry fertilizer application typically shows absorption rates of 20 to 60% (Table 22.1). Fertigation ensures saving in fertilizer (40-60%) due to better fertilizer use efficiency and reduction in leaching.

Table 22.1: Fertilizer use efficiency (%) in fertigation

Nutrient	Soil application	Drip + soil application	Drip + fertigation
N	30-50	65	95
P_2O_5	20	30	45
K_2O	60	60	80

Hypotheses for Fertigation Techniques

1. Fertigation enhances fertilizer use efficiency by 40-60%, hence recommended doses of fertilizers may be reduced proportionally.
2. Drip irrigation promotes root growth in surface layer (about 70-80%), hence the nutrients from sub-surface layers may not be extracted.

3. Drip irrigation leads to moisture content around/above field capacity hence don't promote leaching of nutrients.
4. Use of water soluble fertilizers (WSF) may lead to leaching losses beyond surface layer, hence frequent split application of WSF is desirable.
5. The frequency of fertigation may increase with fertilizers doses in order to avoid leaching losses or toxicity, if any.

Fertigation Scheduling

Factors that affect fertigation module are soil type, available NPK status, organic carbon, soil pH, soil moisture at field capacity, available water capacity range, aggregate size distribution, crop type and its physiological growth stages, discharge variation and uniformity coefficient of drip irrigation system. Fertilizers can be injected into the irrigation system at various frequencies once a day or once every two days or even once or twice a week. The frequency depends upon system design, irrigation scheduling, soil type, nutrients requirement of a crop and the farmers' preference.

The efficient fertigation schedule needs following considerations:

1. Crop and site-specific nutrient management.
2. Timing of nutrient delivery to meet crop needs.
3. Controlling irrigation to minimize leaching of soluble nutrients below the effective root zone.

Methods of Fertilizer Injection

Fertilizers can be injected into drip irrigation system by selecting appropriate equipments.

Commonly used fertigation equipments are:

- Venturi Injector
- Fertilizer Tank (By-pass system)
- Injector Pump
- Automatic Fertigation Controller

1. **Venturi Injector:** Venturi consists of a converging section, a throat and a diverging section. Operational principle of venture is the presence of a constriction in the pipe line (throat section), which accelerate the water flow, reduces the pressure and create a suction effect which is used to pump the fertilizers stock solution into the micro-irrigation system through a tube from the fertilizer storage tank connected to the throat. It allows creation of a

desired pressure difference across the Venturi by using pressure regulating valves to enable the flow of fertilizer stock solution into the water line. Venturi injector usually works on 20% pressure difference from one side of the device to the other. Venturi injector is simple in operation, relatively inexpensive and has good control over the fertilizer concentration with the irrigation water. It is very sensitive to pressure and rate of flow of water in the main line and has relatively low discharge rate. It also causes a high pressure loss in the system which may results in uneven water and fertilizer distribution in the greenhouse.

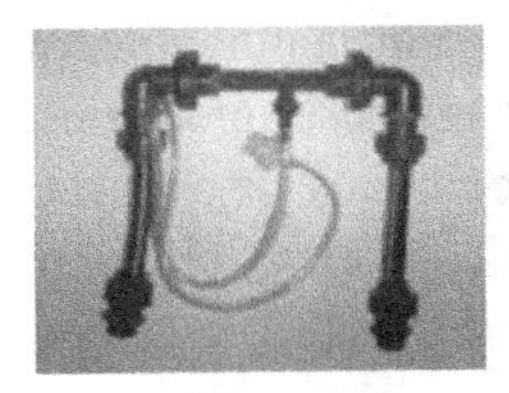

Venturi Injector

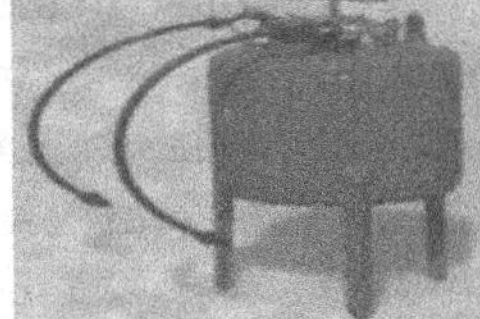

Fertilizer Tank

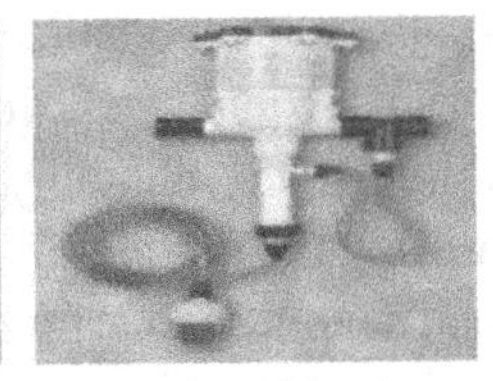

Injector Pump

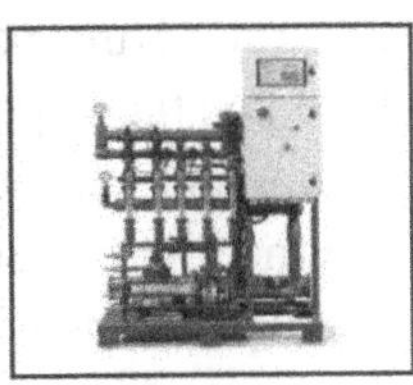

Automatic Fertigation Controller

2. **Fertilizer Tank (Flow by-pass system):** Fertilizer tank is connected to the water line by two small pipes with a pressure reducing valve between the two on the main irrigation line, so a part of irrigation water flows through the fertilizer tank and dilute the fertilizer solution. The duration of fertigation depends on pressure head difference between inlet and outlet and on fertilizer tank size. Fertilizer tank is simple in construction and operation. It works without external energy supply and is relatively lesser sensitive to change in pressure or flow rate. In this system, the concentration of the fertilizer entering the irrigation water changes continuously with the time, starting at high concentration, as a result uniformity of fertilizer distribution can be a problem.
3. **Fertilizer Injector Pump:** Piston pumps and diaphragm pumps are positive displacement pumps, and are recommended where precise control of injection flow rate of fertilizers is required. These pumps usually draw fertilizer stock solution from a storage tank and inject it into irrigation system under hydraulic pressure of the irrigation system. The injection rate of fertilizer solution is proportional to the flow of water in the system and can be adjusted to attain the desired level of fertilizer application. A high degree of control over the fertilizer injection rate is possible, no serious head losses are experienced and operating costs are low. Moreover, if the flow of water stops, fertilizer injection also stops automatically.
4. **Automatic Fertigation Controller:** In greenhouse production, control of greenhouse media-water-nutrient-root environment is critically dependent upon the grower, regardless of whether the operator is a person or computer. Any

disruption or disturbance to the fertigation schedule may quickly create detrimental nutrient stress on the crop. Therefore, control of high frequency fertigation must be automated which is capable of responding to small and rapid changes in nutrient concentration of growing media. Moreover, automated fertigation has reduced production costs in greenhouses and found its way with increased applications in the commercial greenhouse production. It can help in precise nutrient management, irrespective of growing fruits, vegetables, flowers, or medicinal crops. It saves energy, labour and fertilizers costs.

Greenhouse automated fertigation equipment like Fertijet, Fertigal, Fertidoser, Fertimix, Nutricare *etc.* allows fertilizers to be applied to the plants automatically based on their specific nutrient needs during growth cycle. Greenhouse automation equipment can be controlled manually with an on location computer or over the internet thereby limiting the amount of labour. Generally, each and every fertilizer injector unit is usually equipped with anywhere from 3 injectors to 12 injectors. The automated irrigation and fertigation systems can sometimes decrease consumption of water by more than 30% under greenhouse condition.

Steps for Programming Fertigation

- Calculate peak water requirement of a crop.
- Calculate fertilizer dose to be injected according to fertility status and targeted yields.
- Select suitable fertilizers according to solubility, compatibility, nutrient ratio.
- Make a stock solution (1:10), dissolve fertilizers completely in plastic container.
- Check the suction rate of fertigation equipment selected. Do the calibration if required.
- Feed data to computer / controller, if automatic fertigation controller is being used.
- Stop all leakages from micro-irrigation system.
- Operate micro-irrigation system till field capacity in soil is achieved. Follow irrigation schedule.
- Adopt pulse injection fertilizers.
- Adjust pH and EC by adding Nitric acid, Sulphuric acid *etc.* if required.
- Adjust the concentration of fertilizers in effective root zone by adjusting volume / flow of water.
- Check the pH, EC of fertilizer solution coming from last dripper.
- Avoid over irrigation.

Points to Ponder While Adopting Fertigation

- For precise placement of both water and fertilizer, it is necessary to use pressure compensating drippers or inline drippers instead of micro tubes.
- Daily feeding is most desirable, if not possible feed on alternate day.
- Urea can be combined with WSF for meeting the N requirements.
- Fertigation should be done at the end of irrigation period, run the drip system for 5-6 minutes after the completion of fertigation.
- It is not advisable to use a very concentrated stock solution (generally not more than 10%).
- The fertilizer solution should be compatible with the quality of water into which it is being injected.
- Do not inject fertilizers in combination with pesticides or chlorine.
- The injection point must be upstream of the filter system so that the filter will remove any un-dissolved fertilizers or precipitates that occur.
- Select fertilizer solutions to help adjust water pH if necessary.
- The time needed to distribute the fertilizer should be less than the time needed to supply enough water to the field.
- Do not over irrigate because this will lead to leaching of some of the nutrients out of the root zone.

Care to be Taken During Fertigation

- Install micro-irrigation system in greenhouse as per design.
- Stop all leakages from micro-irrigation system.
- Maintain field capacity all the time, do not over irrigate.
- Check the suction rates of fertigation equipments.
- Calculate the correct fertilizer dose to be injected.
- Concentration of fertilizer in effective root zone should be as per the growth stages of crop.
- Use pressure compensating drippers.
- For correct fertigation, fertilizers should be injected in last 40-50 minutes of total irrigation then continue drip irrigation for 5-10 minutes after completion of fertigation.
- Fertilizer doses should be as per soil analysis or targeted yields.

Forbidden Mixture

- Calcium nitrate with any phosphates or sulphates
- Magnesium sulphate with di or mono ammonium phosphate
- Phosphoric acid with iron, zinc, copper and magnesium sulphates

Sources of Nutrients for Greenhouse Crops

Materials used to supply nutrients for greenhouse crop production are chosen based on several factors including cost per unit of nutrients, solubility in water, ability to supply multiple nutrients, freedom from contaminants and ease of handling. The most commonly used fertilizer materials for greenhouse crops are listed in **Table 22.2**. These materials are mostly used to formulate the liquid fertilizer nutrient solution. Pre-mixed fertilizer materials are very popular with many greenhouse operators because they are easy to use. Pre-mixed fertilizers most often supply the P, Mg, S, and micronutrients, and part of the N. They may supply part or all of the K. However, they are relatively expensive compared to individual ingredients and the premixed materials leave little room for making changes in concentrations of individual nutrients.

Table 22.2: Sources of nutrients and their solubility

Fertilizer Grade	N : P : K	Solubility (g/l) at 25 °C
Urea	46:00:00	1100
Ammonium nitrate	34:00:00	1920
Ammonium sulphate	21:00:00	750
Calcium nitrate	16:00:00	1290
Magnesium nitrate	11:00:00	710
Urea ammonium nitrate	32:00:00	1183
Potassium nitrate	13:00:46	133
Mono ammonium phosphate (MAP)	12:61:00	375
Potassium chloride	00:00:60	340
Potassium sulphate	00:00:50	110
Potassium thiosulphate	00:00:25	1550
Mono potassium phosphate (MKP)	00:52:34	230
Phosphoric acid	00:52:00	1906
NPK	19:19:19 or20:20:20	300-400

Recommended Combinations of Water Soluble Fertilizers

Use of two or more fertilizer tanks separates fertilizers that interact and cause precipitation. Follow compatibility chart appropriately to avoid any precipitation in nutrient solution as depicted in **Table 22.3**. Placing in one tank the Ca, Mg and micro-nutrients, and in the other tank, the phosphorus and sulphate compounds enables safe and efficient fertigation.

Clogging of Drip Irrigation System

Clogging in drip irrigation system may be due to the following reasons:

- Activity of algae, fungi, bacteria present in the irrigation water.
- Salt deposition as a result of fertilizer application through the system.
- Presence of sand, silt, organic material *etc*. in the irrigation water.

Trouble Shooting in Micro-irrigation System (MIS)

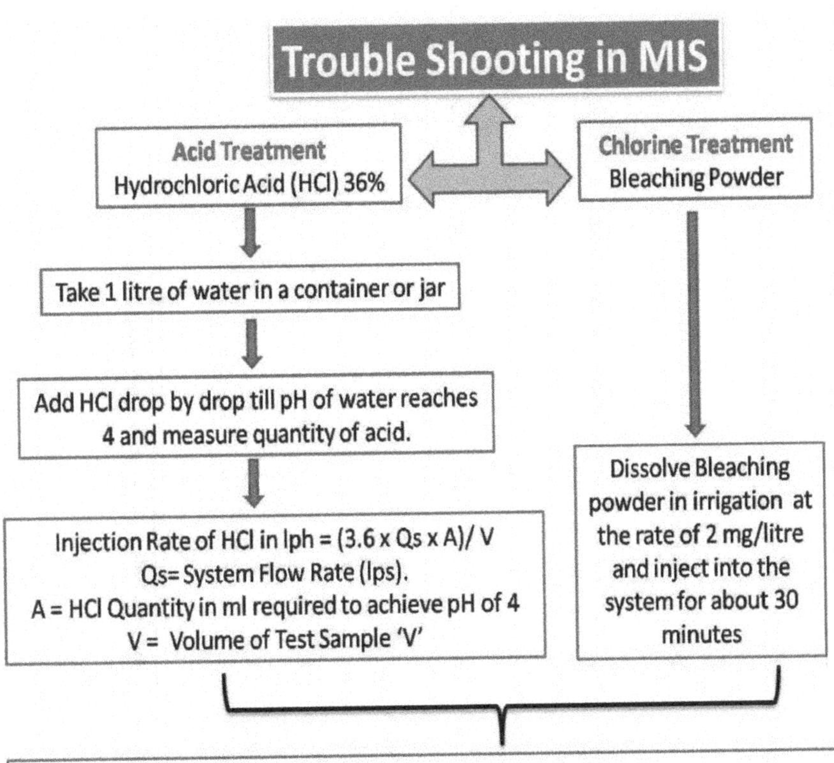

Table 22.3: Fertilizer compatibility chart

	Urea	Ammonium nitrate	Ammonium sulphate	Calcium nitrate	Potassium nitrate	Potassium chloride	Potassium sulphate	Ammonium phosphate	Fe, Zn, Cu, Mn sulphate	Fe, Zn, Cu, Mn chelate	Magnesium sulphate	Phosphoric acid	Sulfuric acid	Nitric acid
Urea	✓													
Ammonium nitrate	✓	✓												
Ammonium sulphate	✓	✓	✓											
Calcium nitrate	✓	✓	x	✓										
Potassium nitrate	✓	✓	✓	✓	✓									
Potassium chloride	✓	✓	✓	✓	✓	✓								
Potassium sulphate	✓	✓	R	x	✓	R	✓							
Ammonium phosphate	✓	✓	✓	x	✓	✓	✓	✓						
Fe, Zn, Cu, Mn sulphate	✓	✓	✓	x	✓	✓	R	X	✓					
Fe, Zn, Cu, Mn chelate	✓	✓	✓	R	✓	✓	✓	R	✓	✓				
Magnesium sulphate	✓	✓	✓	x	✓	✓	R	x	✓	✓	✓			
Phosphoric acid	✓	✓	✓	x	✓	✓	✓	✓	✓	R	✓	✓		
Sulfuric acid	✓	✓	✓	x	✓	✓	R	✓	✓	✓	✓	✓	✓	
Nitric acid	✓	✓	✓	✓	✓	✓	✓	✓	✓	x	✓	✓	✓	✓

✓ = Compatible; x = Incompatible R = Reduced Compatibility

- **Chlorination**: Addition of chlorine is the primary means to control clogging by adding calcium hypochlorite or sodium hypochlorite @ 1-2 mg/lit on continuous basis. Sometimes, chlorine gas is also used to reduce or remove clogging of emitters.
- **Acid treatment**: The chemical salt deposition, which causes clogging can be cleaned by acid treatment. A solution of diluted hydrochloric acid (36 %) of 0.5 to 2 % by volume of water, if introduced in laterals for 10 minutes can effectively remove $CaCo_3$ deposition thus reducing clogging.
- **Pipe line flushing**: Flushing the main, sub-mains and laterals with pressurized water can remove the potential clogging materials like sand, silt, organic/ biological matters present in the system.

Example

Calculation of fertilizer requirement for greenhouse cucumber

N: P: K requirement for a greenhouse cucumber may vary from region to region, however, it is 9.0: 7.5: 7.5 kg/1000 m^2 in this example.

Grower must have clear idea of fertigation scheduling as shown in the table below for greenhouse cucumber.

Crop Duration	Distribution ratio fertilizers			Time of fertigation Initiation	Fertigation frequency
	N	P	K		
First Growth Period (Up to 30 days)	2	3	1	At the appearance of 2nd-true leaf stage	Twice a week
Second Growth Period (30-60 days)	1	2	3		
Third Growth Period (60-90 days)	1	2	3		

Step 1: Make the total of distribution ratio as shown below and allot fertilizers to each ratio as per the recommended dose of NPK (9.0: 7.5: 7.5 kg/1000 m^2).

Crop Duration	Distribution ratio of fertilizers and fertilizer allotment		
	N	P	K
First Growth Period (Up to 30 days)	2 (4.50 kg)	3 (3.20 kg)	1 (1.08 kg)
Second Growth Period (30-60 days)	1(2.25 kg)	2 (2.15 kg)	3 (3.21 kg)
Third Growth Period (60-90 days)	1 (2.25 kg)	2 (2.15 kg)	3 (3.21 kg)
Total of distribution ratio	**4 (9.00 kg)**	**7 (7.50 kg)**	**7 (7.50 kg)**

* Minor adjustments *w.r.t* fertilizer allotment as per ratio may be done to meet out the actual fertilizer requirement of crop.

Step 2: As first growth period of crop is for 30 days and the recommended fertigation frequency is twice a week, which is actually scheduled at the appearance of 2nd-true leaf stage (10 days after sowing). Thus, grower is left with 20 days out of 30 days of initial growth phase in cucumber, so based on the frequency of fertigation and effective days left in the first phenophase of greenhouse cucumber, grower has to perform another calculation as given below:

Crop Duration	**Fertilizer allotment**			**Fertilizer allotment as per Fertigation frequency (twice a week)**		
	N	**P**	**K**	**N**	**P**	**K**
First Growth Period (Up to 30 days)	4.50 kg	3.20 kg	1.08 kg	0.75 kg per fertigation	0.53 kg per fertigation	0.18 kg per fertigation

* In present case, grower is left with 3 weeks (22-21 days) in first growth phase, so there would be 6 fertigations in this phase.

Similarly, grower can make calculations for other growth phase of the crop viz., Second Growth Period (30-60 days) and Third Growth Period (60-90 days)

Step 3: Grower can select any source of water soluble fertilizers enlisted in Table 22.2 with utmost care for prior checking of compatibility of selected fertilizers as per Table 22.3. Further calculation can be done as per the nutrient percent present in the selected fertilizers.

23

PRODUCTION TECHNOLOGY OF CUCUMBER UNDER PROTECTED ENVIRONMENT

Cucumber (*Cucumis sativus* L.) is an edible cucurbit popular throughout the world for its crisp texture and taste. Cucumber is a truly versatile vegetable because of wide range of uses from salads to pickles and digestive aids to beauty products.

SOIL AND CLIMATE

Soil

Cucumbers prefer light textured soils that are well drained, high in organic matter and have a pH of 6-6.8. It is adapted to a wide range of soils, but will produce early in sandy soils. Cucumbers are fairly tolerant to acid soils (pH 5.5). Greenhouse cucumbers generally grow quite well in a wide range of soil pH (5.5-7.5), but a pH of 6.0-6.5 for mineral soils and a pH of 5.0-5.5 for organic soils are generally accepted as optimum. Most of the greenhouse growers use soil-based media for cucumber production. The soil-based medium is composed of 70% red soil, 20% well decomposed organic matter and 10% rice husk. The raised beds of 40 cm height and 90 cm top width are made for successful cultivation of crop. Now-a-days it is an essential practice for all greenhouses to pasteurized soil medium once in two years. However, sometimes necessity may be felt to do so if heavy infection of soil with soil-borne pathogens is observed. This increase in frequency is occasionally necessitated due to the proliferation of disease-causal organisms in the greenhouse. Formaldehyde is a commonly used chemical to sterilize the root medium. Drenching of root medium with formaldehyde (37-40 %) mixed with water in 1:10 proportion at the rate of 7.5 litre for 100 m^2 *i.e.* 37.5 litre of formaldehyde will be required for 500 m^2 of polyhouse.

After drenching, the soil or root medium will be covered with plastic film or black polyethylene sheet. Close all ventilation spaces. Three to four days after formaldehyde treatment remove polyethylene cover. Two days after removing the polyethylene cover rake the bed repeatedly to remove trapped formaldehyde fumes completely before transplanting.

Climate

Air temperature is the main environmental component influencing vegetative growth, flower initiation, fruit growth and fruit quality of greenhouse cucumber. Growth rate of the crop depends on the average 24 hour temperature, higher the average air temperature, faster the growth. The larger the variation in day-night air temperature, taller the plant and smaller the leaf size. Although, maximum growth occurs at a day and night temperature of about 28°C, maximum fruit production is achieved with a night temperature of 19-20°C and a day temperature of 20-22°C. The minimum temperature should not be lower than 18°C for sustained production. Prolonged temperature above 35°C should also be avoided as fruit production and quality are affected at extremely higher temperatures.

Choice of Varieties

Selection of variety is one of the most important decisions made during crop production process. The cucumber plant displays a variety of sex forms like monoecious, gynomonoecious, gynoecious, andromonoecious and androecious; predominantly monoecious type of form is present in most of the varieties. There are also cucumber hybrids that produce fruits without pollination. These varieties are called parthenocarpic varieties resulting in fruits that are called 'seedless', although the fruit often contain soft, white seed coats.

Parthenocarpic cucumber varieties identified by public and private sector for cultivation under protected conditions are listed below:

Public Sector varieties	:	Pant Parthenocarpic Cucumber-2, Pant Parthenocarpic Cucumber-3 (G.B. Pant University of Agriculture and Technology, Pantnagar), Pusa Seedless Cucumber-6 (IARI, New Delhi), KPCH-1 (KAU, Thrissur)
Private Sector varieties	:	KUK-9, 24, 29, 35; Kian, Hilton, Valleystar; Multistar; RS-03602833, Kafka, Oscar, Dinamik etc.

Sowing time and method

Cucumber can be grown successfully round the year under greenhouses. Although, seed is generally sown directly into the soil, but looking to the high cost of seed and problem of competition among the plants during gap filling, It is advisable to raise 5-10% of total population through plug trays so that these can be used for timely gap filling in order to keep pace with the growth of other plants. Generally, 3000 seeds are sufficient for a greenhouse of 1000 m^2. There are three ingredients *viz.*, cocopeat, vermiculite and perlite commonly used as media for nursery raising. These ingredients are mixed in 3:1:1 ratio before filling the trays. Owing to the cost of these ingredients, cocopeat alone can be used as rooting media. The cocopeat usually comes in bricks of 5 kg, but before using it as growing media it must undergo through various hydration processes with water to remove excess of salt present in it. Next step is to hydrate cocopeat brick with calcium nitrate @ 100 g per brick at least for 24 hrs.

Fertilizer Management

Fertigation schedule for cucumber cultivation under greenhouse is given as under:

Fertilizer requirement: 9.0:7.5:7.5 kg NPK/ 1000 m^2

Crop Duration	Distribution pattern/ ratio of fertilizers			Remarks
	N	P	K	
First Growth Period (Up to 30 days)	2	3	1	• Fertigation should be started at the appearance of 2nd true leaf stage.
Second Growth Period (30-60 days)	1	2	3	• Fertigation should be carried out twice a week.
Third Growth Period (60-90 days)	1	2	3	

It is also recommended to apply 0.5 kg *Trichoderma viride*, 0.5 l *Pseudomonas fluorescens*, 2.0 t FYM or 0.4 t vermicompost and 5 kg micro-nutrients (Grade V: Fe 0.2%, Mn 0.5%, Zn 5%, Cu 0.2% and B 0.5%) at the time of sowing.

Irrigation Management

If drippers are placed at 30 cm distance with water discharge rate of 2 lph (litres per hour), adopt the following irrigation schedule for better results.

Crop Stage	Time of operation of drip system (minutes)	Frequency of Irrigation		
		Summer	Winter	Rainy
Upto initiation of first flowering	25	Daily	Alternate Day	Every 4th Day
Fruit Setting to First Harvesting	40	Daily	Alternate Day	Every 4th Day
First Harvest to one week prior to last harvest	35	Daily	Alternate Day	Every 4th Day

Training and Pruning

Cucumber plants in greenhouse are trained to single stem system, which can be achieved by removing all other laterals arising from the axials of leaves, commonly known as suckers at the attainment of 10-12 cm length and only main stem should be allowed to grow vertically along the supporting string.

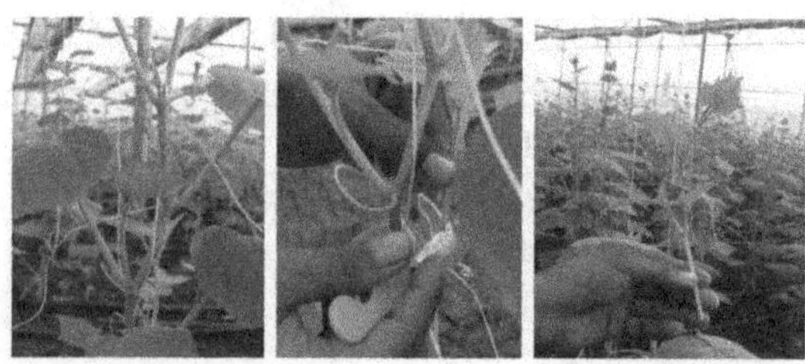

Periodical removal of side shoots to train greenhouse cucumber vertically

Training cucumber plants into Single Stem System

Fruit Thinning

Overbearing can sometimes be a problem. To prevent the plants becoming exhausted and to improve fruit size, control the number of the fruits per plant through selective fruit thinning. This technique is powerful, so use it with great caution. The optimum number of fruits per plant varies with the cultivar and even more with the growing conditions. Although, limiting the number of fruits per plant invariably results in premium priced large fruit. Growers risk underestimating the crop's potential or failing to forecast good weather, they may decide to remove too many fruits and thus unnecessarily limit production. Fruit thinning is undoubtedly most useful in the hands of experienced growers who can use it to maximize their financial returns. Fruit to be pruned must be removed as soon as it can be handled, before it grows too large. This process requires lot of experience to work out exact crop load in cucumber, thus in the lack of exact number of fruits to be thinned, the universal phenomenon of "Survival of the Fittest" may be allowed to persist. But in following this phenomenon, the chances of getting deformed and under sized fruits are intensified significantly.

DISEASE AND PEST MANAGEMENT

Disease Management

Anthracnose (*Colletotrichum orbiculare* & C. *lagenarium*)

- Seeds must be treated with Carbendazim @ 0.25%.
- Field sanitation by burning of crop debris.
- Grow crop on bower system to avoid soil contact.
- Maintain proper drainage in the field.
- Foliar sprays of Carbendazim @ 0.1% or Chlorothalonil @ 0.2% or *Propiconazole* @ 0.1% soon after infection.

Downy Mildew *(Pseudoperonospora cubensis)*

- Crop should be grown with wider spacing in well drained soil.
- Air movement and sunlight exposure helps in checking the disease initiation and development. Bower system of cropping reduces the disease incidence.
- Field sanitation by burning crop debris to reduce the inoculums.
- Protective spray of Mancozeb @ 0.25% at seven days interval gives good control.
- In severe case, one spray of Metalaxyl + Mancozeb @ 0.2% may be given but it should not be repeated.

Powdery Mildew *(Sphaerotheca fuligena and Erysiphe cichoracearum)*

- Foliar sprays of Penconazole @ 0.05% or Carbendazim @ 0.1%.
- Use tolerant line/genotype.

Fruit Rots *(Phytophthora cinnamomi, Pythium, Rhizoctonia, Phomopsis cucurbitae)*

- Avoid soil contact of fruit by using bower system of cultivation and staking of plant.
- Provide proper drainage in the field.
- Green manuring followed by soil application of *Trichoderma* @ 5 kg/ha in soil.
- Collect affected fruits and burn them to reduce primary inoculum.

Gummy Stem Blight *(Didymella bryoniae*-teleomorph *and Phoma cucurbitacearum* anamorph)

- Green manuring followed by *Trichoderma* application @ 5 kg/ha.
- Maintain proper drainage and aeration in the field.
- Seed treatment with Carbendazim @ 0.25%.
- One drenching of Carbendazim @ 0.1% near the collar region.
- Avoid injury near collar region.

Leaf Spots *(Cercospora citrullina, Alternaria cucumerina and Corynespora melonis, Didymella bryoniae* (teleomorph) and *Phoma cucurbitacearum* anamorph)

- Field sanitation, selection of healthy seeds and crop rotation reduces disease incidence.
- Fungicidal sprays of Mancozeb @ 0.25% alternatively with one spray of Hexaconazole @ 0.05%.

Mosaic and Leaf Distortion

- Management of the disease involves destruction of diseased hosts. Virus free seeds must be used to check the seed transmission.
- Initial rouging of the infected plants.
- Periodical spray of systemic insecticides up to flowering stage to control vectors.

Pest Management

Serpentine Leaf Miner *(Liriomyza trifolii* Burgess)

- Soil application of neem cake @ 250 kg/ha immediately after germination.

- Destroy cotyledon leaves with leaf mining at 7 days after germination.
- Spray Pulverized Neem Seed Powder Extract (PNSPE) @ 4% or neem soap @ 1% or neem formulation with 10000 ppm or more (2ml/l) after 15 days of sowing and repeat after 15 days, if necessary.
- If the incidence is high, remove all severely infected leaves first and destroy. Then mix neem soap @ 5 gm and Hostothion @1 ml/l and spray. After one week, spray neem soap 1% or PNSPE or neem formulation with 10000 ppm or more (2ml/l).
- Never spray the same insecticide repeatedly.

Red Spider Mite *(Tetranychus neocaledonicus* Andre)

1. Remove and destroy the affected leaves.
2. Spray neem oil/neem soap/ pongamia soap 1%.
3. Spray need-based application of acaricides like Abamectin 1.8% EC @ 2 ml/ 10 L or Milbemectin 1% EC @ 5 ml/ 10Lor Spiromesifen 22.9% EC @ 8 ml/ 10 L or Fenazaquin 10 EC @ 1 ml/L in rotation with plant products like pongamia oil or neem oil (8-10 ml/l) or neem soap (10 g/l).
4. When incidence is severe, remove and destroy all severely infected leaves followed by a spray of mixture of an acaricide with botanicals mentioned above.

Thrips *(Thrips palmi Karny)*

- Soil application of neem cake (once immediately after germination and again at flowering) followed by NSPE @ 4% and neem soap 1% alternately at 10-15 days interval.
- Spray any systemic insecticides like Acephate 75 SP @ lg/l or Dimethoate 30 EC @ 2ml/l.

Leaf Eating Caterpillar

- Apply neem cake to soil immediately after germination.
- Spray any contact insecticides like Carbaryl 50 WP @ 3g/l. Neem or pongamia soap @ 0.75% also effectively manages this pest.
- Soil application of neem cake (once immediately after germination and again at flowering) followed by NSPE @ 4% and neem soap 1% alternately at 10-15 days interval.
- Spray Carbaryl 50 WP @ 3g/l or Indoxacarb 0.5 ml/l.

Standard operating procedures (SOPs) for nematode control in green house

Nematodes such as *Meloidogyne incognita, M. javanica* (root-knot nematodes), *Rotylenchulus reniformis* (reniform nematode), *Heterodera sp.*, (cyst nematodes) are important nematode species attacking plants.

The following can be adopted for securing healthy seedlings/crops:

Step 1. Construction of new polyhouse/nethouse

Soil samples collected from proposed sites for the construction of new polyhouses/ net houses should compulsorily be tested for nematode infestation.

Step 2. Raising nematode-free planting material

Select nematode free planting materials. Seed treatment with bio-pesticides- *Pseudomonas fluorescens* @ 10g/kg seed.

Step 3. Removal of roots from previous crops

Roots harbour nematode eggs, and each root gall contains hundreds of nematode eggs. Infected roots of previous crop should be removed as much as possible. This single practice can remove 80-90% of nematode inoculum from the soil.

Step 4. Preparation of bed

Soil fumigation with Formalin@ 75ml/m^2 or Metham sodium @ 30ml/m^2 + neem cake @ 200g/m^2 enriched with *Trichoderma* / *Pseudomonas*/ *Purpureocillium lilacinum* or *Pochonia chlamydosporia*@ 50g/m^2 15 days prior to sowing.

Step 5. Soil Solarization

During peak summer (May-June), after the crop is over soil solarization should be carried out with thin (25 ìm) transparent polythene sheet.

Step 6. Organic amendments fortified with bio-agents

Once the solarization process is over, remove the polythene sheets from polyhouse, prepare the beds and mix the bio-agent (*Trichoderma*, *Pseudomonas*, *Purpureocillium lilacinum* and *Pochonia chlamydosporia*) fortified FYM uniformly over the beds in top soil. Neem cake powder @ 50-100 g/m^2 of planting bed may be mixed on the top layer about 7-10 days before seeding/transplanting. Nursery bed treatment with *Trichoderma harzianum* @ 50 g/ sq. m.

Step 7. Crop Rotation/ trap cropping

Adopt suitable crop rotation practices with non host crops. Growing cowpea closely to the main crop and removing the plants from polyhouse at 45 days after sowing. Marigold can also use as antagonistic crop for nematode management.

Step 8. Usze of Nematicides

Carbofuran 3 G @ 10g/m^2 can be applied in soil at the time of seeding/transplanting. Soil application of Dazomet @ 40g/m^2 before 15 days of sowing.

Yield

Generally, cucumber is ready for first harvesting in 35 to 40 days of planting depending upon climatic conditions and crop management practices. Harvesting is done when fruits are more or less cylindrical and well filled and should be carried out in early morning or late evening. The produce should immediately be moved to cool, shaded and ventilated area. As fruits are harvested manually, so these should be clipped or snapped with a slight twist motion and should not be pulled off the vines to minimize 'pulled ends". Yield of 10-15 t/1000 m^2 can be obtained from greenhouse cucumber.

24

PRODUCTION TECHNOLOGY OF BELL PEPPER UNDER PROTECTED ENVIRONMENT

Capsicum (*Capsicum annuum* L. var. *grossum* Sendt.) commonly known as sweet pepper /bell pepper or *Shimla mirch*, is a member of the family Solanaceae and considered as luxury vegetable being used in pizza making. It is gaining popularity among farmers due to its quick and high returns. Under open conditions, quality and productivity of the produce is poor, which reduces profit margins of capsicum growers. Bell pepper has attained a status of high value crop in India in recent years and occupies a pride of place among vegetables in Indian cuisine because of its delicacy and pleasant flavour coupled with rich content of ascorbic acid and other vitamins and minerals. Capsicum growing in India is of recent interest due to high demand for fast food dishes in hotels and modern restaurants.

SOIL AND CLIMATE

Soil

Bell pepper grows best on loam or sandy loam soil with good water holding capacity. However, it can be grown on a variety of soils as long as they are well drained. For successful pepper cultivation, soil pH should be in between 5.5 to 6.8. Soil media may encounter problems of soil-borne pathogens, so it is important to draw representative samples for checking the status of soil-borne pathogens as well nutrient status prior to planning bell pepper cultivation inside the protected structures. Depending upon the analysis report in relation to the presence of soil-borne pathogens, growers can adopt any of the soil pasteurization methods (Refers to chapter 19) for effective control of these pathogens.

Climate

Bell requires relatively cool weather and the day temperature less than 30°C is favourable for growth and yield. Higher temperature results in rapid plant growth but it affects fruit set. Lower night temperature (20°C) favour flowering and better fruit set. Shading and misting are essentially required during summer to avoid temperature build up inside greenhouses. Moderately high RH (50-60%) is preferred, which can be maintained by ventilation and foggers.

Choice of Varieties

Selection of greenhouse sweet pepper cultivars is dependent on colour, disease resistance, performance and yield. A variety of seed companies and distributors offer greenhouse sweet pepper cultivars and the latest cultivars are always subject to change when superior cultivars are developed. The cultural requirements of the different cultivars can be distinct enough to require that the environments be managed differently in order to obtain maximum yield. The following list of bell pepper varieties in different colour segments shows wide range of selection suited appropriately to region specific needs. However, bell pepper growers must be fully convinced about the variety likely to be taken up for cultivation.

Sr. No.	Colour segment	Varieties
1.	Green	Indra, Lario, NS-631, NS-632, NS-274, NS-292, NS-626, Bharat, Mahabharat, Indam K1, Indam K2, Green Queen, Master, Indam Mumtaz
2.	Red	Bomby, Inspiration, Pasarella, NS-280, NS-631, NS-632, King Anther, Nun- 3010, Hira, Laxmi, Mamtha, Inspiration, Pasarella
3.	Yellow	Orobelle, Bachata, NS-281, NS-626, Yellow Queen, Nun-3020, Tanvi, Golden Summer, Super Gold, Swarna, Bachata,
4.	White	White-1
5.	Chocolate	Chocolate Wonder
6.	Orange	Sympathy

Sowing Time, Method and Spacing

Bell pepper can successfully be grown round the year under greenhouses through proper management practices. Bell pepper nursery is raised on soil-less media *viz.*, cocopeat, vermiculite and perlite in pro trays having 98 plugs. Seed @ 40 g is sufficient for an area of $1000m^2$. Seedlings are planted at the spacing of 45 x 30 cm in green segment while pacing for coloured bell pepper is 45 x 45 cm. But, care must be taken while transplanting the seedling of capsicum to avoid the direct contact of collar region of the seedlings to soil.

Fertilizer Management

Fertigation schedule for capsicum cultivation under greenhouse is given as under:

Fertilizer requirement: 25:25:25 kg NPK/ 1000 m^2

Crop Duration	Distribution ratio of nutrients			Time of fertigation Initiation	Fertigation frequency
	N	P	K		
1st Growth Period (Up to 30 days)	2	3	1	10-15 days of planting	Twice a week
2nd Growth Period (31-60 days)	1	2	2		
3rd Growth Period (61-90 days)	1	1	3		
4th Growth Period (91-120 days)	1	1	2		
5th Growth Period (121-150 days)	1	1	1		
6th Growth Period (161-180 days)	1	1	1		

It is also recommended to apply 0.5 kg *Trichoderma viride*, Phosphorous Solubilizing Bacteria (*Bacillus megaterium*), Azotobacter, *Pseudomonas fluorescens* each, 0.4t vermicompost and 5.0 kg micro-nutrients (Grade-5) per 1000 m^2 at the time of planting.

Irrigation Management

If drippers are placed at 30 cm distance with water discharge rate of 2 lph (litres per hour), adopt the following irrigation schedule for better results.

Crop Stage	Operational Time of MIS (minutes)	Frequency of Irrigation		
		Summer	Winter	Rainy
Upto fruit Setting	60-75	Alternate Day	Every 4th Day	Every 4th Day
Fruit setting to first harvesting	75-90	Alternate Day	Every 4th Day	Every 4th Day
First harvest to one week prior to last harvest	60-75	Alternate Day	Every 4th Day	Every 4th Day

Training and Pruning

The growing point at the top of the plant is removed after 20-25 days of plating, which is known as is called topping. This technique is adopted for producing more branches. Capsicum plants can be trained into various systems *viz*., two stem system, three stem system and four stem system. In two stem system, two main stems are maintained on each plant after pruning or pinching the other branches by leaving two leaves or at least one leaf and one flower at each internode. Accordingly, plants can be trained into three or four stems by maintaining three or four stems. At every node, growing tip splits into two giving rise to one strong branch and one weak branch and the weaker one is removed. These stems are trained upon strings to the main overhead wire at 8-10 feet height. The

Pinching of apical bud after 20-25 days of planting for emergence of side shoots

Training bell pepper (green segment) to 4 shoot system for higher yield

stems are either loosely trellised or bound around the strings. The stems may be clipped with strings using rings or plastic clips. The bell pepper plants can grow up to 8-9 feet height in 9 to 10 months period and after that the plants are topped to avoid stem breaking and to improve fruit size. **The most appropriate system for green bell pepper grown for consumption in local market is four shoot system, whereas coloured bell pepper should be trained in two shoot system for getting appropriate fruit size acceptable at market.**

Fruit Thinning

When there are too many fruits on the plant, it is necessary to remove some fruits to promote the development of remaining fruits. Fruit thinning is done when the fruit is of pea size. This practice is normally followed to increase the size of fruit thereby increasing the quality of production.

DISEASE AND PEST MANAGEMENT

Disease Management:

Dieback and Anthracnose *(Choanephora capsici, Colletotrichum capsici)*

Management

- Disease free seeds collected from healthy fruits should be sown.
- Seeds should be treated with Carbendazim @ 0.25% prior to sowing.
- Seedling should be sprayed by Carbendazim @ 0.1% before transplanting. Cut the rotting twigs along with healthy part and burn it.
- Foliar spray of Copper Oxychloride @ 0.3% followed by Carbendazim @ 0.1% at flowering stage. Avoid apical injury during transplanting and also at flowering stage.

Bacterial Leaf Spot

Management

- Seed dipping in Streptocycline solution @ 100 ppm for 30 minutes.
- One spray of Streptocycline @ 150 ppm alternatively with Kasugamycin @ 0.2%.

White Rot *(Sclerotinia sclerotiorum)*

Management

- Cut the infected plant parts along with some healthy portion in morning and carefully collect in polythene to avoid falling of sclerotia in the field. Burn all these materials away from field.
- Foliar spray of Carbendazim @ 0.1% at flowering stage followed by Mancozeb @ 0.25%.

Leaf Blight *(Alternaria alternata and Cercospora capsici)*

Management

- One foliar spray of Chlorothalonil @ 0.2% alternatively with Thiophenate-methyl@ 0.1% after 8-10 days.
- Selection and sowing of disease free seeds to check the primary infection.
- Field sanitation by burning of infected crops debris followed by summer ploughing.

Phytophthora Leaf blight/Fruit Rot

Management

- Always use healthy seeds.
- Infected crop debris and fruits must be collected from the field and burnt.
- Preventive sprays of Mancozeb@ 0.25%.
- One spray of Metalaxyl+ Mancozeb @ 0.2% is very effective when applied within two days of infection but repetitive sprays should not be given.
- Proper staking of plant reduces disease infection. Follow crop rotation, proper water management and drainage for lesser incidence. Avoid high nitrogen application.

Leaf Curl Complex

Management

- Sow healthy seeds.
- Seed treatment with hot water at 50^{0}C or 10% tri sodium phosphate solution for 25 minutes.
- Nursery should be raised inside insect proof net to check the vector infestation.

- Root dipping of the seedlings in Imidacloprid solution @ 4-5 ml per litre of water for one hour prior to transplanting.
- Periodical alternate spray of *Spiromesifen* @ 1 ml/ 1 followed with *Wettable Sulphur* @ 0.2%.
- Initial rouging of infected plants soon after infection and burn it.

Sun Scald (Physiological disorder)

Management

- Adequate fertility and proper water management will help to develop the canopy of leaves and foliage required to protect the fruit from sun scald.
- Timely shading of plants by closing the top shade nets in the greenhouse.
- Poor foliage cover allows the defect to occur. Variety selection may play a significant role; compact plants may not provide cover as that of more vigorous plants.

Insect-Pest Management

White fly

- Apply neem cake @ 250 kg/ha to plant beds while planting and repeat after 30 days.
- Spray Acephate 75 SP@ 1.0g/l or Fipronil 5 SC @ 1ml/L or Ethofenprox 10 EC @ 1ml/l in rotation.
- Spray Acephate 75 SP (0.5 g) + pongamia oil (2 ml) +1 ml sticker in one litre water after emulsifying (shaking thoroughly in a bottle).

White or Yellow Mites *(Polyphagotarsonemus latus)*

- Apply wettable sulphur 80 WP @ 3g/l or any acaricide (directing the spray on the ventral surface of leaves).
- Spray pongamia oil (2ml/l) mixed with acaricides. Spray neem seed powder extract 4% at 10 days interval when the pest incidence is low.
- Spray Abamectin 1.9 EC @ 0.5ml/l or Spiromesifen @ 1 ml/l or Fenazaquin 10 EC@ 1ml/l in rotation with plant products like pongamia oil or neem oil (8-10 ml/l) or neem soap (10g/l)..

Aphids *(Aphis gossypi* and *Myzus persicae)*

- Spray Acephate 75 SP @ 1g/l or Dimethoate 30 EC @ 2 ml/l in rotation when required.
- Remove all the virus affected plants and destroy immediately.

Borers (Tobacco caterpillar, *Spodoptera litura* and *H. armigera*)

- Spray specific NPV of the borer species. Inundative release of *Trichogramma* sp. eggs.
- Use marigold as trap crop (one row of marigold for every 18 rows of bell pepper) for managing *H. armigera.* Collect and destroy eggs masses and immature larvae of S. *litura.*
- Use poison baiting (10 kg rice flour + 1kg of jaggery + 250 g of Thiodicarb) for *S. litura* and repeat the baiting 2-3 times, if necessary.
- For *S. exigua,* spray Indoxacarb 14.5 SC @ 0.75 ml/l or Spinosad @ 0.75 ml/l or Thiodicarb 75 WP@ 0.75 g/l. Sometimes tomato fruit borer and tobacco caterpillar may also attack capsicum under polyhouse. Follow the management practices given under tomato (Chapter 25).

Root-knot Nematodes

- Follow Standard operating procedures for the management of nematodes (Chapter 23).

Yield

Harvesting of capsicum is done at green, breaker and coloured (red/ yellow *etc.*) stage. It depends upon the purpose for which it is grown and distance for the ultimate market. In India fruits are harvested at breaker stage for long distant markets. For local market, it is better to harvest at coloured stage. Breaker stage is the one when 10% of the fruit surface is coloured and when more than 90% of the fruit surface is coloured it is considered as coloured stage.

Harvesting of capsicum fruits starts 60-70 days after transplanting and continues up to 170 to 180 days at 7 to 10 days interval. Mature green or fruits with 10-15% colour development weighing around 150 to 250 g are harvested. Keep the harvested fruits in cool place and avoid direct exposure to sunlight. All damaged, malformed and bruised fruits should be removed. Those with dirt adhering to their surface can be cleaned by wiping the surface with a moist soft cloth. The bell pepper fruits should be graded into same size and colour lots according to market requirements. Yield of 10 to 12 t/1000 m^2 (10 to 12 kg/m^2, 2.25 to 2.70 kg/plant) can be expected from one crop of 8-9 months duration.

The capsicum fruits can be categorized into following grades:

Sr. No.	Grade	Fruit weight
1	A+	> 200 g
2	A	150-200 g
3	B	100-150 g
4	C	< 100 g

25

PRODUCTION TECHNOLOGY OF TOMATO UNDER PROTECTED ENVIRONMENT

Development of varieties/ hybrids both from public and private sector has significantly played important role in enhancing crop production. Greenhouse tomato production has attracted much atten-tion in recent years, partly because of a new wave of inter-est in alternative crops. The attraction is based on the per-ception that greenhouse tomatoes are more profitable than the more conventional horticultural crops.

Soil

Tomatoes can be produced across a wide range of soils as long as drainage and physical soil structure is good. Optimum soil pH is between 6.0-6.5, but crops can thrive well in soils with a pH of 5.0-7.5. When pH drops below 5.5, the availability of magnesium and molybdenum goes down and above 6.5, zinc, manganese and iron become deficient. Follow the standard procedure for soil sampling for the analysis of nutrients and soil-borne pathogens and adopt management strategies if presence of soil-borne pathogens is ascertained (Follow any standard method motioned in chapter 19).

Climate

Temperature is the primary factor influencing all stages of development of tomato from vegetative growth to flowering, fruit setting and then fruit ripening. Growth requires temperature between 10°C and 30°C. Temperature and light intensity affect vegetative growth, fruit set, pigmentation and nutritive value of fruits. The minimum temperature for germination of seeds ranges from 8 to 10°C. The night temperature is the critical factor in fruit setting with optimum range of 16°C to

22°C. Fruits fail to set at 12°C or below. Fruit set is also reduced markedly when average maximum day temperature goes above 32°C and average minimum night temperature goes above 22°C. However, now some varieties are available which can set fruit beyond the critical temperature limits and are called hot set (set fruits at higher temperature) and cold set cultivars (set fruits at lower temperature). Light intensity is one of the major factors affecting the amounts of sugars produced in leaves during the photosynthesis, and this, in turn, affects the number of fruits that the plant can support and the total yield. Optimum relative humidity in glasshouse crops range from 60 to 70%. Under hydroponics, 75% and 85% respectively are typical night and daytime relative humidity. The humidity beyond this range may results in very poor fruit setting in tomato.

Choice of Varieties

Selection of the most suitable cultivar is a pre-requisite for successful tomato cultivation in a greenhouse. The greenhouse cultivars are indeterminate in growth habit and plants may reach to a length of 10 feet or even more during the growing period. The first step in raising any crop is to choose the best variety. There are thousands of tomato varieties available on the market, but only a few are acceptable for greenhouse pro-duction. These are exclu-sively indeterminate varieties/hybrids bred specifically for greenhouse production.

Criteria for the Selection of Tomato Variety (s) for Greenhouse Cultivation

- Size of fruit desired.
- Disease resistance.
- Lack of physiological problems, *i.e.,* cracking, cat facing, blossom end rot.
- Yield uniformity.
- Market demand.

Tomatoes grown in the greenhouses are generally divided into different categories as mentioned below:

1. **Beefsteak cultivars:**The cultivars of the beefsteak category produce large slicing type fruits weighing around 180 to 250g. The fruits are generally harvested individually and usually packed with calyx attached. These cultivars are very popular among greenhouse growers in almost all the European countries and USA.
2. **Big fruited cultivars:** The fruits of such cultivars usually weigh around 80-150g and come in small to medium clusters. Although, several cultivars have

been developed for greenhouse cultivation in Europe, USA, Turkey, Israel *etc.*, but in India, there is a limitation of cultivars bred exclusively for protected cultivation. Otherwise, indeterminate cultivars from public as well private sector are generally cultivated under protected conditions, list of which is given below:

Indeterminate/ Semi-determinate Hybrids/ Varieties in India

Public Sector	Pant Polyhouse Tomato-1, Pant Polyhouse Tomato Hybrid-1, Azad T-5, COTH 1, COTH 2, TNAU Tomato Hybrid CO3, Punjab Gaurav, Punjab Sartaj, Him Palam Tomato Hybrid-1, Pusa Ruby, Pusa Uphar, Pusa Hybrid-2, Arka Abhijit, Arka Ahuti, Arka Ananya, Arka Shreshta, Arka Rakshak, Arka Samrat
Private Sector	Anup, Avinash-2, Trishul, Heemsohna , To-848, Rakshita, Naveen 2000+, ARTH-128, Avtar, Naveen, NTH-2004, NTH-2005, NTH-2008, Vishwas, Saberano, NS-4266

3. **Hand type or cluster type:** These cultivars produce fruits in clusters of four to seven or even more and are generally harvested and marketed with full cluster having fruits from breaker to ripe stage. The weight of single fruit of these cultivars is around 50-70g.
4. **Cherry tomato:** This category is getting plenty of attention from growers as consumers are now-a-day preferring such cultivars for table purposes. These are very small in size, may be round or oval in shape and average fruit weight is 12-20g depending upon the cultivar. Punjab Red Cherry, Punjab Sona Cherry (yellow coloured fruits), Punjab Kesar Cherry(Orange coloured fruits) [PAU, Ludhina]; Solan Red Round (UHF, Solan); Phule Jayshree (MPKV, Rahuri); Gujarat Anand Cherry Tomato 1 (AAU, Anand); Pusa Cherry Tomato 1 (IARI, New Delhi); Swarna Ratan (ICAR-RCER, Ranchi); NS574 (904), NS575 (907), NS 6438 (Namdhari Seeds Pvt. Ltd.); Laila, Sheeja, Roja, Ruhi (Known you seed India Pvt. Ltd.) are some of the cherry tomato cultivars identified by public and private organizations.
5. **Coloured tomato:** More recently, the cultivars with outstanding flavoured and rich in antioxidants and vitamin A are gaining importance among the greenhouse growers for better price. In this category, brown and yellow coloured cultivars are available for cultivation.

Sowing Time, Nursery Raising and Spacing

Tomato can also be grown throughout the year under greenhouse conditions but special attention has to pay for off-season production. Nursery for greenhouse tomato is raised under protected structure, mostly in soil-less media in plug trays to produce disease free and mainly virus free seedlings. 30 g seed is sufficient for planting in an area of 1000 m^2. The ingredients of soil less-media *viz.*, cocopeat, vermiculite and perlite are mixed in 3:1:1 ratio prior to filling of pro trays or plug trays. Generally, plug trays with 98/104 plugs or small plugs are preferred for sowing of tomato seeds.

Fertilizer Management

The subject of fertility is probably among the most con-fusing for greenhouse tomato growers. The keys to a successful nutrition program include the following:

- Use fertilizer designed specifically for greenhouse tomatoes.
- Know how much of each fertilizer element is needed.
- Know how much is being applied.
- Check the electrical conductivity (EC) and pH levels.
- Be observant for signs that plants may be deficient or have an excess of a nutrient.
- If possible, monitor plant nutrient status by periodically taking samples for tissue analysis.

The crop is fertigated with N: P: K at the rate of 25:12.5:12.5 kg per 1000 m^2. The application of 0.5 kg *Trichoderma viride* and *Pseudomonas fluorescens* each, 0.5 L Phosphorous Solubilizing Bacteria (*Bacillus megaterium*) & potash mobilizer- *Frateuria aurantia* each, 2 t FYM and 5.0 kg micro-nutrients (Grade V) should also be carried out at the time of transplanting

Crop Duration	Application ratio of fertilizers			Time of fertigation Initiation	Fertigation frequency
	N	P	K		
1st Growth Period (Up to 30 days)	2	3	1	10-15 days of planting	Twice a week
2nd Growth Period (31-60 days)	1	2	2		
3rd Growth Period (61-90 days)	1	1	3		
4th Growth Period (91-120 days)	1	1	2		
5th Growth Period (121-150 days)	1	1	1		
6th Growth Period (151-180 days)	1	1	1		
7th Growth Period (181-210 days)	1	1	1		

Irrigation management

When drippers are placed at 30 cm distance with water discharge rate of 2 lph (litres per hour), adopt the following irrigation schedule for better results.

Crop Stage	Time of operation of drip system (minutes)	Frequency of Irrigation		
		Summer	Winter	Rainy
Upto fruit Setting	60	Alternate Day	Every 4th Day	Every 4th Day
Fruit Setting to first harvesting	75	Alternate Day	Every 4th Day	Every 4th Day
First harvest to one week prior to last harvest	60	Alternate Day	Every 4th Day	Every 4th Day

Training and pruning

For best production, prune tomato plants to a single stem by removing all lateral shoots, commonly referred to as 'suckers'. One sucker will form at the point where each leaf originates from the main stem, just above the leaf petiole (stem). Allowing all suckers to grow and bear fruit would increase the total number of fruit, but they would be small and of poor quality. It is better to have one main stem (s) that bears fruit, as this will produce larger, more uniform, and higher quality fruit. Removing suckers once in a week will keep them under control. It is advisable to leave one or two of the smallest suckers at the top of the plant. Then, if the plant becomes damaged and the terminal breaks off, one of these suckers can be allowed to grow and become the new terminal. Generally, remove any sucker longer than 3 cm.

Plants are trained vertically along the supporting wire to explore the full potential of indeterminate varieties. When the plant reaches to the overhead wire, it should be leaned and dropped. Lower all plants to the same height so they don't shade each other. Repeat this operation each time plants grow higher than the wire. Always twin the plants in the same direction, if you start clockwise, continue clockwise; otherwise, when the plant gets heavy with fruit, it may slip down the string and break. Some growers prefer to use plastic clips to secure the plant to the string, either in combination with wrapping or to replace wrapping.

Cluster pruning will also improve size and uniformity. This involves removing small fruit from some clusters, leav-ing three, four or five of the best ones. Remove misshaped or deformed fruit first, otherwise, remove the smallest fruit, which is usually the last one formed on each cluster.

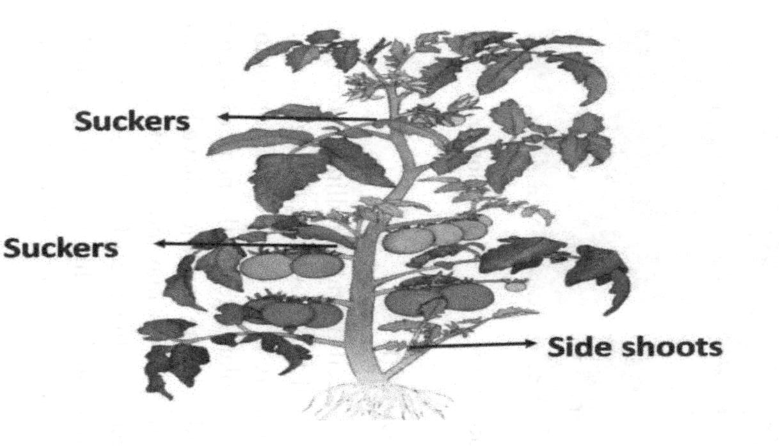

Emergence of side shoots or suckers in tomato

Removal of suckers in tomato to train in single stem system

Pollination

Tomato flowers bear both male and female organs within a flower (hermaphrodite flowers). Botanically, these are termed as perfect flow-ers. In the green-house, wind is not strong enough to shake the flowers suffi-ciently to transfer the pollen. Even though the greenhouse is ventilated with fans or on cooler days when the fans are not operating, the air is relatively motionless.

The optimum temperature for pollination should within the range 21 to 27 °C. Optimum relative humidity is 60%, so at RH more than 70%, pollen grains stick together and are not dispersed properly leading to poor fruit set inside the greenhouse environment. With relative humidity less than 60% for extended periods, the stigma may dry out and pollen grains will not stick to it. Greenhouse tomato growers can use an electric pollinator to ensure good fruit set.

Electric pollinator

Cell operated pollinator

Blower

Poolination Tools for Greenhouse Tomato

Vibrate tomato truss with electric pollinator or cell operated pollinator for 5-10 seconds on every 3rd day starting from the day of first flowering during morning hours when stigma receptivity and dehiscence are at their peak. This operation can also be performed manually by shaking the plants gently the peak of stigma receptivity and dehiscence in the morning and growing conditions have humidity at permissible limit. However, these tools sometimes may not be effective at higher humidity levels, so growers can resort to the use of different growth hormones which can facilitate fruit setting under such conditions (very high RH). Different growth hormones along with their role in tomato production are listed below.

Use of plant growth regulators for fruit in tomato:

PGRs	Common name	Dose (ppm)	Effectiveness	Remarks
2-Chloroethyl-phosponic acid	Ethephon	200-500	Flowering induction and setting	· Foliar spray at 50% Flowering and repeat it at 20-25 days interval
2-Chloroethyl (trimethyl ammonium chloride)	Cycocel	500-1000	Flower bud stimulation and increase in fruit set	
2,4-Dichlorophenoxy acetic acid	2,4-D	2-5	Increase fruit set, earliness and parthenocarpy	
3- Indole Butyric acid	IBA	50-100	Increase fruit set	
Para-chlorophenoxy acetic acid	PCPA	30-50	Fruit set under adverse climatic condition	· Foliar spray at 50% Flowering and repeat it at 20-25 days interval
4-Cholorophenoxy Acetic Acid	4-CPAA	30-50	Increase fruit set, parthenocarpy	· Truss dipping at every 3rd or 4th day under very high humid environment for excellent fruit set. Appropriate dose for truss dipping is 30ppm

Disease and Pest Management

Damping-off (*Pythium, Rhizoctonia, Phytophthora, Fusarium* etc.)

1. Hot water treatment of seeds (at 52°C for 20 min.).
2. Treat the seeds with Captan /Metalaxyl-Mancozeb @ 2.5-3.0 g/kg of seeds.
3. Drenching the plug trays 7 days before sowing with Thiram/Captan or any copper fungicide @ 3g / litre of water.
4. Remove the affected seedlings from trays as soon as the symptoms are visible.

Early blight (*Alternaria solani, A. alternata f. sp. lycopersici*)

1. Crop rotation with non-solanaceous host is essential for effective reduction of inoculum load.
2. Field sanitation by plucking the lower leaves and burning of infected crop debris. Minimize relative humidity around plant canopy for preventing the infection.
3. Spray the crop with 0.3% Mancozeb or 0.1% Carbendazim or Chlorothalonil 0.2% at 10-15 days interval starting from 45 days after transplanting as a prophylactic measure.
4. Two spray of Chlorothalonil @ 0.2% or Propiconazole (0.1%) at 8 days interval is effective against the disease but spray must be started soon after infection on floral parts.

Late blight (*Phytophthora infestans* (Mont) de Bary)

1. Always use healthy seeds.
2. Infected crop debris and fruits must be collected and burnt.
3. Preventive sprays of Mancozeb @ 0.25% provide good control during cloudy and cold weather but spray interval should be 5 to 7 days.
4. One spray of Metalaxyl+ Mancozeb @0.2% is very effective when applied within two days of infection but repetitive sprays should not be given.
5. If late blight is present in pockets, start spraying with Metalaxyl-Mancozeb (0.2%) / Cymoxanil-Mancozeb (0.25) / Dimethomorph (0.1%) + Mancozeb (0.2%) immediately and alternate spray schedule with protectant and systemic fungicides at 7-8 days interval.
6. Training an pruning of plants should be performed timely.

Powdery mildew (*Leveillula taurica*)

1. Spray with Karathane (0.1%) or Wettable Sulphur (3 g/ litre of water) twice at an interval of 10 days.

Bacterial Spot *(Xanthomonas campestrispv. Vesicatoria)*

1. Summer ploughing to desiccate the bacteria and host.
2. Plug tray nursery raising to avoid seedling infection.
3. Seedlings dip in streptocycline solution @ 100 ppm.
4. One spray of Streptocycline @ 150-200 ppm followed by one spray of Copper Oxychloride @ 0.2%.
5. One spray of copper oxychloride @ 0.3% after 15 days of antibiotic application.

Bacterial wilt (*Psendomonas solanacearum*)

- Treat the seeds with hot water at 52°C for 20 minutes or with 0.01% streptomycin solution for 30 minutes.
- Follow proper crop rotation with non-solanaceaous crops for at least 2 to 3 years.
- Apply bleaching powder @ 15-20 kg/ha and 4.5q lime in soil at least 3-4 weeks before planting.
- Proper drainage should be maintained.

Leaf Curl Complex (Virus-transmitted by white fly as well as by mechanical injury)

1. Removal of weed host from nearby surrounding areas or from inside polyhouse if present.
2. Seed treatment with hot water at 50°C or trisodium phosphate solution (10%) for 25 minutes.
3. Cover the nursery area with 40 mesh insect screen and spray the seedlings with Imidacloprid (3.5 ml/ 10 L).
4. Seedlings dip in Imidacloprid @ 4-5 ml per litre of water for one hour prior to transplanting.
5. Plant barrier crops of taller non-host crops like maize, bajra and sorghum outside the polyhouse.
6. Apply Carbofuran 3G @ 25-30 kg/ha 10 days after transplanting followed by 2-3 foliar sprays with 0.05% Dimethoate (1.5 ml/L) or Imidachlorpid (3ml/ 10L) or Thiomethoxam (3.5 ml /10L) at 10 days interval and alternate the spray schedule with NSKE (0.5%).

7. Periodical sprays of systemic insecticides up to flower setting.
8. Avoid mechanical injury during intercultural operations.
9. Use tolerant varieties.
10. Roughing of infected plants soon after infection at initial stage of growth.
11. Avoid cultural practices like pruning, training etc. with the hands using tobacco and its products; otherwise go for disinfection of hands either with whey or milk powder solution.

INSECT-PESTS

Serpentine Leaf Miner *(Liriomyza trifolii* Burgess)

Management

1. Often the incidence starts from nursery itself. Hence, remove infected leaves at the time of planting or within a week of transplanting.
2. Plastic mulching may prevent pupation in soil.
3. Apply neem cake in beds @ 250 kg/ha at time of planting and repeat after 25 days.
4. Spray neem seed powder extract 4% or neem soap 1% at 15-20 DATP (days after transplanting).
5. If the incidence is high, remove infected leaves and spray Azadirachtin 300ppm @ 3ml/1 L.
6. Avoid frequent spraying of synthetic pesticides. At the most, one spray of Deltamethrin 2.8 EC @ 1ml/L or Cypermethrin 25 EC @ 0.5 ml/1 l may be given, if required.

Greenhouse white fly (*Trialeurodes vaporariorum*)

Management

1. Use virus resistant hybrids, if available.
2. Raise nursery in plug trays inside insect proof nets.
3. Spray Imidacloprid 200 SL (0.3ml/L) or Thiomethoxam 25 WP (0.3 g/L) in nursery at 15 days after sowing.
4. Remove the leaf curl infested plants as soon as disease symptoms are expressed. This helps in reducing source of inoculums of the disease.
5. Seedlings dip in Imidacloprid 200 SL (0.03ml/L) or Thiomethoxam 25 WP (0.3 g/L) before transplanting.

6. Spray Imidacloprid 20 SL (0.3ml/L) or Thiomethoxam 25 WP (0.3 g/L) at 15 days after planting.
7. Install yellow sticky traps coated with adhesive or sticky glue at crop canopy level for monitoring adult white fly population.
8. If the traps indicate the whitefly activity, spray Dimethoate 30 EC @ 2ml/L or neem seed kernel extract 4% (NSKE) or pongamia or neem oil (8-10 ml/L) or neem soap (10 ml/L).
9. Rogue out the virus affected plants as soon as the symptoms are observed.

Fruit Borer *(Helicoverpa armigera* Hubner)

Management

1. Spray *Ha* NPV at 250 LE/ha + 1% jaggery along with sticker (0.5 ml/L) during evening hours when the larvae are young.
2. Spay Indoxacarb 14.5 SC @ 0.5 ml/l or Thiodicarb 75 WP @ 1g/L for grown up larvae.
3. Use Helilure @
4. Spray *Chlorantraniprole 18.5 SC @ 0.3 ml/10 L.*

Tobacco Caterpillar *(Spodoptera litura)*

Management

1. Collection and destruction of egg masses and gregarious larvae.
2. Spray *Spodoptera* NPV 250 LE/ha + 1% jaggery along with sticker (0.5 ml/L) during evenings.
3. Use poison baiting: Mix 10 kg of rice bran or wheat bran with 2 kg jaggery by adding a little water in the morning. In the evening add 250 gm of Thiodicarb formulation and sprinkle over the bed. Caterpillars get attracted to fermenting jaggery, feed and get killed.

Red Spider Mites *(Tetranychus urticae)*

Management

1. Remove and destroy the affected leaves.
2. Spray neem oil/neem soap/ pongamia soap @1%.
3. Spray Abamectin 1.8% EC @ 2 ml/10L or Milbemectin 1% EC @ 5 ml/10L or Spiromesifen 22.9% EC @ 8 ml/10L or Fenazaquin 10 EC @ 1 ml/L in rotation with plant products like pongamia oil or neem oil (8-10 ml/L) or neem soap (10 ml/l).

4. When incidence is severe, remove and destroy all severely infected leaves followed by a spray of mixture of an acaricide with botanicals mentioned above.

Root-knot Nematodes

- *Follow Standard operating procedures for the management of nematodes given in chapter 23.*

IPM PACKAGE FOR INSECT-PESTS IN TOMATO UNDER POLYHOUSE

Raise seedlings in pro trays inside insect proof net of 40 mesh size	
15 DAS (Days after seed sowing)	Spray seedlings with Imidacloprid or Thiamethoxam
One day before transplanting	Drench seedlings with Imidacloprid or Thiamethoxam
At transplanting	Apply neem cake 250 kg/ha.
15 DATP (days after transplanting)	Spray the seedlings with Imidacloprid or ThiomethoxamInstall yellow sticky traps to monitor whitefly.
25 DATP (days after transplanting)	Apply neem cake 250 kg/ha
Post flowering and fruiting stage	Monitor insect-pests like fruit borer, tobacco caterpillar, leaf miner, whitefly and red spider mite. Spray NPV according to the pest.
	Remove leaves, severely infected with leaf miner/ red spider mite
	Spray neem seed powder/neem soap for leaf miner.
	Spray acaricide/botanicals in rotation to control red spider mite, spray systemic insecticide/botanical to control whitefly.

Yield and Storage Management

Harvesting of tomato fruits is a continual process throughout the growing season. Generally, most of the varieties are ready for first picking in 70-75 days after transplanting. The fruits should be harvested preferably early in the morning or late in the evening to avoid post harvest losses and are then graded, packed according to grades. Cherry tomatoes are mostly harvested with stems attached or sometimes singly with calyx attached with the fruits and are packed in the containers of 400-500g capacity. On the whole, 25-30 tonne of big fruited tomatoes and 10-15 tonne of cherry tomatoes can be harvested from 1000 m^2 greenhouse cultivated area.

26

PRODUCTION TECHNOLOGY FOR MELONS UNDER PROTECTED ENVIRONMENT

Muskmelon and watermelon are the important cucurbitaceous vegetable crops grown over a large area in different parts of India and are high in demand during hot months. They are dessert fruits mainly eaten for sweetness, pleasant flavour and also have thirst quenching ability. The normal crop of melons is grown through seed sowing from end of February to second week of March and ready for harvesting in the month of May-June. But the crop can be grown under greenhouses for off-season production to obtain very high prices of the produce.

Climate and Soil Requirement

Melons can be grown in hot and dry atmosphere but plants are sensitive to low temperature and frost. A humid climate may favour the development of foliar diseases. High humidity and excess moisture at the time of fruit maturity may hamper the quality of fruits. The optimum temperature for plant growth is 28-30°C. Comparatively low humidity and high day temperature during ripening period with enough sunshine are conducive for the development of flavour and total soluble solids (TSS) in fruits. These conditions are also suitable for reducing the chances of foliar diseases.

A well drained sandy loam soil is most suitable for the cultivation of melons. Melons are slightly tolerant to soil acidity and prefer a soil pH of 6.0 to 7.0.

Selection of variety

Musk melon	Water melon
Varieties having netted fruits are mostly grown under greenhouse conditions in Israel and other countries for off-season cultivation and export to high markets	Varieties with medium sized fruits are preferred.
Majority of varieties bears andromonoecious flowers except Pusa Rasraj (Monoecious flower), thus requires artificial pollination for fruit set	All varieties bear monoecious flowers, also requires artificial pollination for fruit set.

List of the open pollinated varieties identified/released in India by public Sector

Crop	National level	State level
Muskmelon	Kashi Madhu, Pusa Sarbati, Hara Madhu, Pusa, MHY-5, Madhuras, Arka Rajhans, Arka Jeet, Durgapura Madhu, NDM-15, Pusa Rasraj (F_1)	Punjab Sunehari, Punjab Rasila, Arka Rajhans, Hisar Madhur, RM-43, MHY-3, RM-50, Kashi Madhu **Hybrids:** Punjab Hybrid-1, MHY-3, MHL-10, DMH-4
Watermelon	Durgapura Meetha, Sugar Baby, Arka Manik, Arka Jyoti (F_1)	Durgapura Kesar, Durgapura Lal RHRWH-12 (F_1)

List of cultivars/ hybrids identified by Private Sector for general cultivation

Crop	Name of cultivars/ hybrids
Musk melon	NS-915, NS- 89, NS- 931, NS-972, Dipti, Madhuras, Madhurima, Urvashi, Madhulika, MHC-5, MHC-6, DMH-4, Bobby,
Water melon	Nutan, Madhubala, MAdhuri, Aashtha, Mithasnina, Khushboo, NS-701, NS-702, NS-705, NS-200, Madhu, Milan Nath- 101, MHW- 4, 5, 6, 11, Mohini, Amrit, Hanoey, Suman- 235, Netravati

Planting

Under greenhouse conditions, melons are usually planted as off-season crop. Seeds of melons can be planted directly on the beds. However, in case of costlier seeds, first of all seedlings should be raised in plug trays or polyethylene tubes under protected conditions. Plants get ready for transplanting in 25-20 days. Melons are planted in paired rows on each bed at plant spacing of 60 cm x 45 or 60 cm x 50 cm.

Crop	Seed Rate (per 1000 m^2)
Muskmelon	100-150 g
Watermelon	200-250 g

Training and Pruning

Melon plants are trained vertically inside the protected structures along a training string (preferably UV stabilized) to climb up the overhead wires. The base of the twine is anchored loosely to the base of the stem with a non-slip noose. Removal of secondary shoot upto the seventh node is found optimal in muskmelon to improve plant growth and fruit set and induce early flowering.

Training muskmelon vertically inside greenhouse

Training watermelon vertically inside greenhouse

The side branches are pruned up to the plant height of 45-60 cm in water melon. The side branches are pruned only after leaving one or two pistillate flowers. Plants are trained vertically without any damage to branches as well as pistillate flowers.

Use of Plant growth regulators (PGR) for early appearance of female flowers

Crop	PGR	Dose	Stage of Application
Muskmelon	Ethrel	250 ppm	First spray at 2-true leaf stage Second at 4-true leaf stage
Watermelon	2,3,5- Triiodobenzoic Acid (TIBA)	25-50 ppm	-do-

Pollination

Both the melons require artificial pollination for better set owing to their flower structure *i.e.* andromonoecious in muskmelon and monoecious in watermelon. One colony of honey bees *(Apis melifera)* having 20000 bees is sufficient for effective pollination in 1000 m^2 area of greenhouse. The direction of the beehive and proper ventilation are important factors for efficient working of the bees. Sometimes, honey bees may not respond well to greenhouse environment, so hand pollination can also be employed to affect artificial pollination. As stigma in both the crops is receptive in early morning, so hand pollination must be attempted during peak hours of stigma receptivity. Maximum of 3 fruits are retained per vine in muskmelon while watermelon plants support the crop load of 2 fruits per plant only.

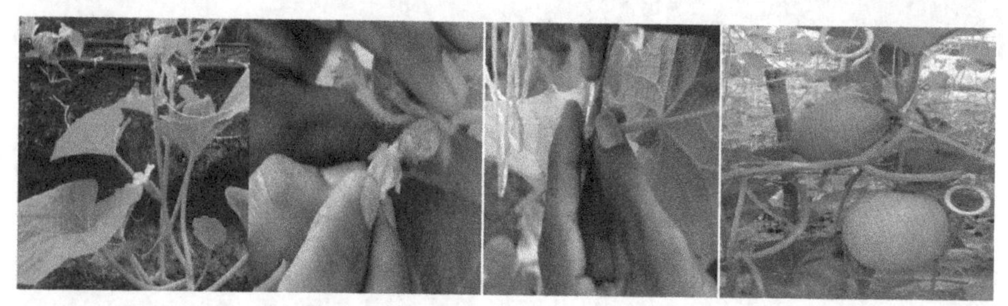

Hand polination in muskmelon **Fruit support**

Fruit Support

Small, mesh bags (onion sacks), cheesecloth or nylons can be used as slings to support the fruit. The bags can be tied to the trellis or the support wire. The bag should allow light penetration and not hold moisture. These support bags can be removed from the trellis upon ripening of the fruits.

Hand polination in watermelon

Fruit support

Fertigation

Fertigation schedule for melon cultivation under greenhouse is given as under:

Fertilizer requirement: 25:20:30 kg NPK/ 1000 m^2

Crop Duration	Distribution pattern / ratio of fertilizers			Remarks
	N	P	K	
First Growth Period(Up to 30 days)	2	3	1	• Fertigation should be started at the appearance of 2nd-true leaf stage. • Fertigation should be carried out twice a week.
Second Growth Period(30-60 days)	1	2	2	
Third Growth Period(60 onwards)	1	1	3	

It is also recommended to apply 0.4 t vermicompost and 5 kg micro-nutrients (Grade-5) at the time of planting.

Irrigation Management

When drippers are placed at 30 cm distance with water discharge rate of 2 lph (litres per hour), adopt the following irrigation schedule for better results.

Crop Stage (Muskmelon)	Time of operation of drip system (minutes)	Irrigation Frequency
Upto 30 days	25	Alternate Day
31 to 60 days	40	Alternate Day
After 60 days	35	Alternate Day

Crop Stage (Watermelon)	Time of operation of drip system (minutes)	Irrigation Frequency
Upto initiation of flowering	30	Alternate Day
Fruit Setting to First Harvesting	50	Alternate Day
First Harvest to one week prior to last harvest	40	Alternate Day

Physiological Disorders in Muskmelon

A) **Fruit cracking:** Fruit cracking in muskmelon occur due to boron deficiency. Spraying of Boron @ 50 g /25 L water should be done to control fruit cracking. Fertigation with 50 g boron should be also done at weekly interval.

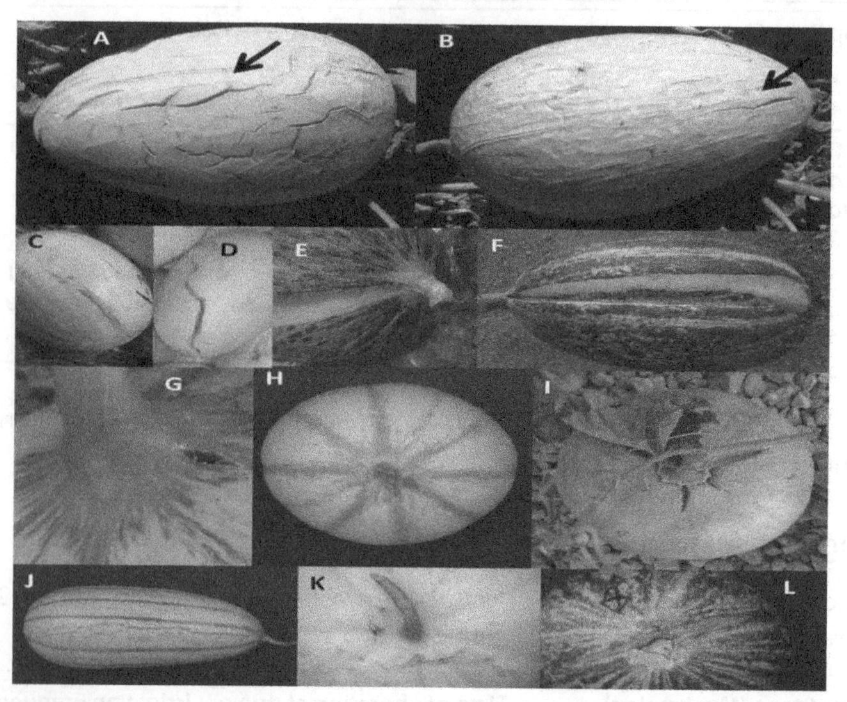

Melon splitting symptoms. (A, B) The black arrow is the typical net cracking with deep furrow as seen in Cartagena, Spain; (C) Transportation splitting; (D) Splitting in packing house; (E) Splitting following in part the netting cracks; (F) Splitting with signs of direct sun exposure; (G) Start of cracking; (H) Cracking around the peduncle immediately after harvest; (I) Splitting in senescent fruit; (J) Cracking associated to full maturity; (K) Cracking; (L) Blossom-end.

B) **Fruit drop:** Fruit drop occur due to improper/poor pollination. Care should be taken during hand pollination.

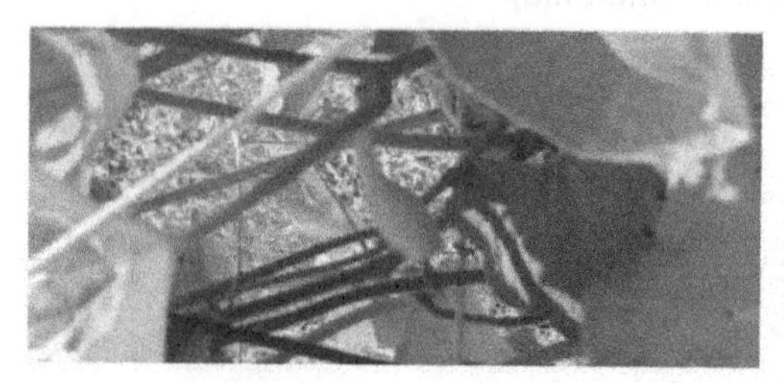

Fruit drop in muskmelon

Physiological Disorder in Watermelon

1) **Blossom End Rot (BER):** Blossom End Rot (BER) is a physiological or non-parasitic disorder related to calcium deficiency, moisture stress or sometimes both. Preventive measures include adequate application of calcium, maintenance of desirable soil pH (6 to 6.5), and a uniform and sufficient supply of water. The incidence of BER is quite variable and it tends to occur more readily in oblong melons.
2) **Hollow heart and white heart:** Hollow heart and white heart (HH & WH) are two physiological disorders influenced by genetics, environment and probably by a number of nutritional factors. To decrease the incidence of these two problems, only cultivars tolerant to HH or WH should be planted. In addition, the crop should be grown under optimal nutritional and moisture conditions.
3) **Sunscald:** Sunscald is damage to the melons caused by intense sunlight. Sunscald can be particularly severe on dark colored melons. Developing and maintaining adequate canopy cover to afford protection (shade) to the melons may prevent sunscald. Sunscald reduces quality by making melons less attractive and may predispose the melon to rot.
4) **Stem Splitting:** Stem Splitting can occur in seedlings being raised for transplanting. This problem seems to be associated with high humidity and moisture that can occur under greenhouse conditions. Watering evenly to maintain soil moisture, avoiding wet dry cycles in the media and good air circulation may help alleviate this problem.

Blossom-End Rot   Hollow heart

Sun scald   Stem splitting

Plant Protection

Diseases	Management Practices
Anthracnose *(Colletotrichum orbiculare & C. lagenarium)*	• Seed should always be collected from healthy fruits and disease free area. • Seeds must be treated with Carbendazim @ 0.25%. • Field sanitation by burning of crop debris. • Train plants vertically to avoid soil contact. • Maintain proper drainage in the field. • Foliar sprays of Carbendazim @ 0.1% or Chlorothalonil @ 0.2%.
Downy Mildew *(Pseudoperonospora cubensis)*	• Crop should be grown with wide spacing in well drained soil. • Air movement and sunlight exposure helps in checking the disease initiation and development. • Proper training and pruning help in reducing the disease incidence. • Field sanitation by burning crop debris to reduce the inoculums. • Protective spray with Mancozeb @ 0.25% at seven days interval. • Under severe incidence, one spray of Metalaxyl + Mancozeb @ 0.2% may be given but it should not be repeated.
Powdery Mildew (*Sphaerotheca fuligenaand Erysiphe cichoracearum*)	• Foliar sprays of Penconazole @ 0.05% or Carbendazim @ 0.1%. • Use tolerant variety, if available.
Gummy Stem Blight (*Didymella bryoniae* teleomorph *and Phoma cucurbitacearum*)	• Avoid exotic hybrids and varieties due to high degree of susceptibility. • Summer ploughing and green manuring followed by *Trichoderma* application. • Maintain proper drainage and aeration in the field. • Seed treatment with Carbendazim @ 0.25%. • Drenching with Carbendazim @ 0.1% near the collar region. • Avoid injury near collar region.
Leaf Spots (*Cercospora citrullina, Alternariacucumerina and Corynespora melonis, Didymella bryoniae* (teleomorph) and *Phoma cucurbitacearum*)	• Field sanitation, selection of healthy seeds and crop rotation reduces disease incidence. • Sprays of Mancozeb @ 0.25% alternatively with Hexaconazole @ 0.05%.

[Table Contd.

Contd. Table]

Diseases	Management Practices
Mosaic and Leaf Distortion	• Destry diseased hosts and weeds. • Virus free seeds must be used to check the seed transmission. • Initial rouging of the infected plants. • Periodical spray of systemic insecticides up to flowering stage to control vectors.
Serpentine Leaf Miner (*Liriomyza trifolii* Burgess)	• Soil application of neem cake @ 250 kg/ha during field preparation. • Destroy cotyledon leaves with leaf mining at 7 days after germination. • Spray PNSPE @ 4% or neem soap 1% or neem formulation with 10000 ppm or more (2ml/L) after 15 days of sowing and repeat at 15 days interval, if necessary. • If the incidence is high, first remove all severely infected leaves and destroy. Then spray a mix of neem soap @ 5 gm and Hostothion @ 1 ml/L. After one week, spray neem soap 1% or PNSPE or neem formulation with 10000 ppm or more (2ml/L). • Never spray the same insecticide repeatedly.
Red Spider Mite (*Tetranychusneocaledonicus* Andre)	• Spray neem or pongamia soap at 1% on lower surface thoroughly. • Alternately, spray Spiromesifen @ 1ml/l or Fenazaquin 10 EC @ 1 ml/L.
Thrips (*Thrips palmi Karny*)	• Soil application of neem cake (once during field preparation and again at flowering) followed by NSPE spray @ 4% and neem soap @ 1% alternately at 10-15 days interval. • Spray any systemic insecticides like Acephate 75 SP @ l g/L or Dimethoate 30 EC @ 2ml/L.
Leaf Eating Caterpillars	• Soil application of neem cake (once during field preparation and again at flowering) followed by NSPE spray @ 4% and neem soap @ 1% alternately at 10-15 days interval. • Spray Indoxacarb 0.5 ml/L.
Root-knot Nematodes (*Meloidogyne incognita)*	• Follow Standard operating procedures for the management of nematodes (Chapter 23)

Harvesting

Maturity Indices

Crop	Days to first harvest	Maturity indices/stage	Yield per 1000 m^2
Muskmelon	75-80	When the fruits slip easily from the vine. Fruit should show changes in colour and degree of netting, and a softening at the blossom end. Best harvest maturity follows in one to three days; and best flavor is attained if musk melons are held near 21°C for this final ripening.	6-8 tonne
Watermelon	90-110	Fruits are harvested when they attain full size. At this stage the curly tendril closest to the point of fruit attachment is often shriveled or dried.	18-20 tonne

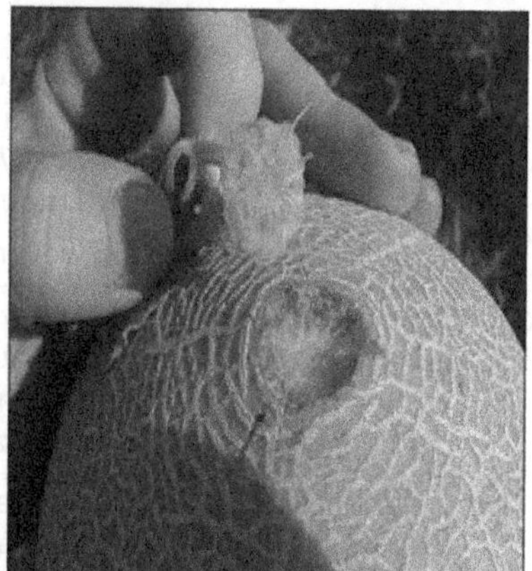

Maturity Index in muskmelon

Maturity Index in watermelon

27

PRUNING: A VIRTUOUS AND UNCONVENTIONAL ALTERNATIVE FOR MASS MULTIPLICATION OF GREENHOUSE CUCUMBER AND TOMATO

Cucumber and tomato are very important and valuable greenhouse vegetable crops. The availability of predominantly monoecious type cultivars/ hybrids in cucumber necessitates artificial pollination, if grown under greenhouse conditions. This requires special attention during chemical control of insect-pests and diseases when natural pollinators are placed inside greenhouse structures and sometimes greenhouse environment is not suitable for pollinators' activities owing to high humidity inside, thus leading to insufficient pollination and poor yield in monoecious type of varieties. However, development of parthenocarpic cucumbers has revolutionized greenhouse cucumber production because of presence of pistillate flowers only setting fruits without pollination and fertilization, resultantly higher yield per unit area and better quality of the produce. It was Noll who introduced the term parthenocarpy and studied it first time in cucumber. This trait is proved to be highly useful to develop fruits under environmental conditions that are unfavourable for successful pollination and fertilization, particularly in greenhouse cultivation. In case of tomato, selection of indertminate varieties is an essential component for successful greenhouse cultivation.

The concept of growing cucumber and tomato under greenhouse production system always stipulates training of plants to maximize vertical space utilization under such system. Greenhouse cucumber and tomato plants are generally trained into single stem system through timely and continuous pruning of side shoots or laterals/ or suckers throughout the growing season. Main purpose of training and pruning is to allow maximum interception of light and better air circulation for proper growth and development of plants and is also helpful to reduce risk of fungus and insect problems. So, these laterals/ suckers can be utilized as planting material in those areas where relay cropping is possible or farmers have numbers of commercial units. Propagation through these cutting can also be useful for staggered planting to make the availability of commercial products throughout the year. Sometimes, unavailability of seeds of hybrids at peak time poses lot of problems and farmers are left with no choice. Vegetative propagation through cutting not only provides true to type plants but also ensures the timely availability of the same for cultivation.

The development of gynoecious varieties with strong parthenocarpic expression is not only the major challenge to cucumber breeders for parthenocarpic hybrid development in lieu of higher yield, earliness, uniformity and suitability for protected cultivation, but making available seeds of such hybrids to greenhouse growers at reasonable rates in developing nations like India. Parthenocarpic varieties do not require pollination, and are therefore less vulnerable to poor pollination conditions under greenhouse environment and the need for insect pollinators.

It is worth to mention here that unlawful exploitation of varieties is transgression under the current epoch of intellectual property. As India is a signatory member of World Trade Organization (WTO) and International Union for the Protection of New Varieties of Plants(UPOV), which demands protection of plant varieties to encourage plant breeders by granting new plant variety a status of 'intellectual property'. In this context, Government of India enacted an act on "Protection of Plant Variety and Farmers' Rights" adopting an effective *sui generis* system to provide a balance between plant breeders' rights along with farmers' rights and researchers' rights. The PPV & FR Act, 2001 safeguards the interest of Indian farmers through exemptions which entitle a farmer to produce, save, use, sow, exchange, share or sell his farm produce including seed/planting material of a variety protected under this Act as a reward considering farmers' past, present and future contributions in conserving, improving and making available plant genetic resources.

PLANT PRODUCTION PROCESS FROM PRUNED SHOOTS

Selection of Rooting Media

It has been observed that soil-less media consisting of cocopeat, vermiculite and perlite in 3:1:1 ratio on volume basis is best for raising of these cuttings. But, owing to the high cost of vermiculite and perlite, it is rather advisable at farmer level to select cocopeat as rooting media. It has also been critically noticed that healthy planting materials can be raised in coco-peat alone, which is as good as raised in three constituents of rooting media. Cocopeat from Indian subcontinent contain several macro and micro-plant nutrients, including substantial quantities of potassium, sodium and chloride. So, before using coco-peat as rooting media, it is always very important to wash cocopeat with quality water to remove excess of elements which are soluble in water such as potassium, sodium and chloride. Washing of cocopeat helps to reduce the electrical conductivity to a tolerance limit. Thereafter, buffering of cocopeat is done with calcium nitrate @ 100 g per 10 litres of water for 5 kg of cocopeat. During this process, calcium [2^+] is introduced in order to remove monovalent positive ions such as potassium [1^+] from the coconut complex. In this way, the process not only removes the soluble elements but also those elements which are bound to the coconut complex. Ideally, the treated water ($CaNo_3$) should be administered into cocopeat over a 24 hours period via a slow splinker system if possible at farmer level, otherwise cocopeat can be rested in calcium nitrate solution for 24 hours. Once the resting period is over, cocopeat is rinsed with water twice and then, cocopeat becomes ready to use as rooting medium. Now-a-days, buffered cocopeat is also available in the market, which can be used directly without following the above mentioned process of washing and buffering.

The laterals shoots/ suckers are then planted in the media preferably in the evening hours. The plug trays having 50 and 98 or 104 plugs are good for cucumber and tomato, respectively.

Methodology

- The side shoots/ laterals of 8-10 cm should be taken from each plant for raising true to true type plants in cocopeat with utmost care of not taking cuttings from virus affected plants (if any).
- The whole process should be performed in the afternoon at 4.00 pm onwards for better survival of the cuttings, when the mist chamber facilities are not available. However, the process of taking cuttings can be performed at any time of the day if one has mist chamber facilities. These side shoots are immediately transferred to plug trays having cocopeat as growing media.
- These plug trays are then put in put in a water bath covered with polyethylene sheet from the top and side to initiate healing process.
- Relative humidly of more than 85% should be maintained for survival.
- Once side shoot cuttings start to show rooting, reduce water supply till completion of rooting process.

- After the completion of rooting process, withheld water supply and let the rooted cuttings get hardened. In this way, the transplants get ready for planting in 15-20 days.

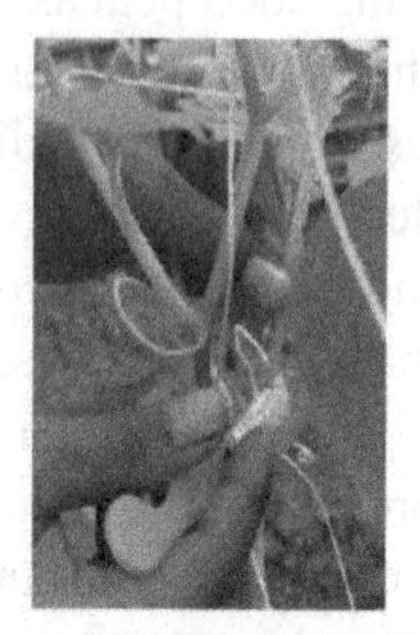

Removal of side shoots as a part of training cucumber into Single Stem and their utilization for plant multiplication

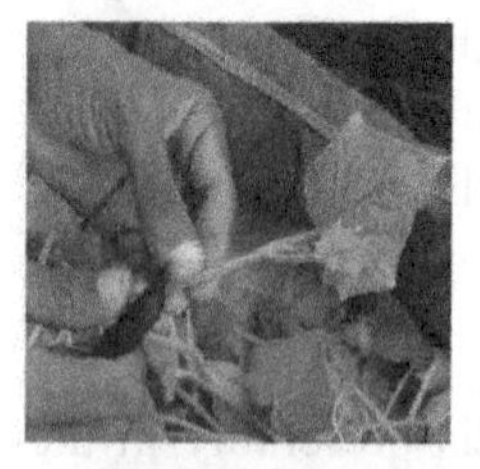
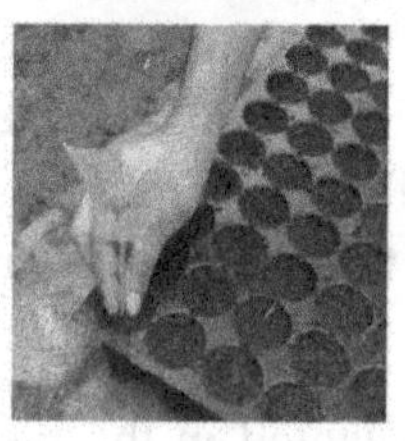

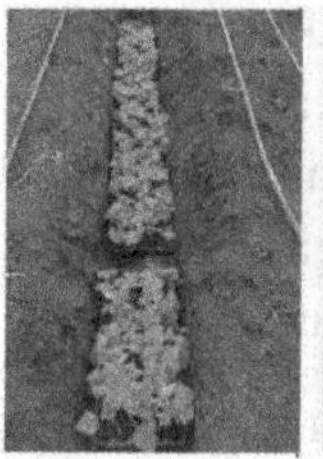

Trasferring cuttings onto plug trays and water bath for healing and rooting

Rooting in cucumber transplants raised from side shoot cuttings

Cucumber crop raised through side shoot cuttings

Removal of suckers as a part of training tomato on Single Stem and their utilization for plant multiplication

Rooting in transplants and crop raised from suckers in tomato

Conclusion

Multiplication of greenhouse cucumber and tomato cultivars through side shoot cuttings opens up new dimensions in greenhouse production system. Greenhouse farmers can grow successive crops of cucumber and tomato by generating true to type plants through side shoot cuttings, which otherwise goes as waste in the process of training plants vertically. The crops of cucumber and tomato raised either through seeds or side shoot cuttings perform equally well for various desired horticultural traits. The farmers may even sell excessive plants generated through side shoot cuttings for additional income. Handful amount of greenhouse cucumber and tomato cultivars available with farmers can be used to generate plenty number of plants required for cultivation in a larger area of greenhouse. In this way, the burden of high cost of seeds can be minimized to a greater extent.

28

SYSTEMS OF SOIL-LESS CULTIVATION

Soil-less culture is defined as the method of cultivating crops without using soil as a rooting medium, in which nutrients to growing crop are supplied through the irrigation water. The fertilizers containing nutrients are dissolved in water at an appropriate concentration to formulate a solution called nutrient solution, which is supplied to maintain proper health of the crop grown in the system.

The rapid expansion of soil-less system all across the globe over a period of last 3-4 decades may be attributed to the independence of growing system from the soil and related problems like soil-borne pathogens and the decline of soil structure and fertility due to continual cultivation with the same or relative crop plants. Soil-less cultivation is proposed as the safest and most effective alternative to soil sterilization by chemical methods. It is therefore becoming increasingly important and popular system of cultivation for protected culture not only in modern fully automated greenhouses, but also in simple greenhouses like naturally ventilated polyhouses (NVPH). Soil-less culture offer numerous advantages:

1. Absence of soil-borne pathogens.
2. Safe alternative to chemical sterilization of soil.
3. Possibility to cultivate greenhouse crops and achieve high yields and good quality even in saline or sodic soils, or in non-arable soils with poor structure.
4. Precise control of nutrition particularly in crops grown on inert substrates or in pure nutrient solution (also in soil-less crops grown in chemically active growing media, plant nutrition can be better controlled than in soil grown crops due to the limited media volume per plant and the homogeneous media constitution).
5. Avoidance of soil tillage and preparation thereby increasing crop length and total yield in greenhouses.
6. Enhancement of early yield in crops planted during the cold season because of higher temperatures in the root zone during the day.

7. Minimum ecological impact (*e.g.* reduction of fertilizer application and restriction or elimination of nutrient leaching from greenhouses to the environment). Therefore, application of closed hydroponic systems in greenhouses is compulsory by legislation in few countries particularly in environmentally protected areas or those with limited water resources.

Despite the considerable advantages of commercial soil-less culture, there are disadvantages limiting its expansion in some cases:

1. Demands higher initial capital investment for installation of the system.
2. Requires high technical knowhow to execute crop cultivation in such systems.

Soil-less culture (Hydroponics) is a broad term used to describe all the techniques of growing plants in aerated nutrient solution (water culture) or in solid media other than soil (substrate culture).

Soil-less cultures usually have higher yields than soil culture systems. An optimum quality of growing substrate is one of the most important components involved in successful production, because it directly affects the availability of water and nutrients to the plant. Materials such as aged pine bark, coconut pith, compost, peat moss, perlite, sand, vermiculite, and wood by products (*e.g.* sawdust, composted pine bark) can potentially be used as growing medium components for horticultural production systems. The two broad categories of soil-less culture are given below:

Classification of Soil-less Culture Systems

Soil-less culture	
Water culture or hydroponic	**Substrate culture**
• Deepwater culture	• Gravel culture
• Float hydroponic	• Sand culture
• Nutrient film technique	• Bag culture
• Deep flow technique	• Container culture
• Aeroponics	• Trough culture

Hydroponic system are further classified as open system when the nutrient solution is delivered to plants and is not reused, and closed system when the surplus is recovered, replenished and recycled. Since regulating the aerial and root environment is a major concern in such agricultural systems, production takes place inside enclosures designed to control air and root temperatures, light, water, plant nutrition *etc.*

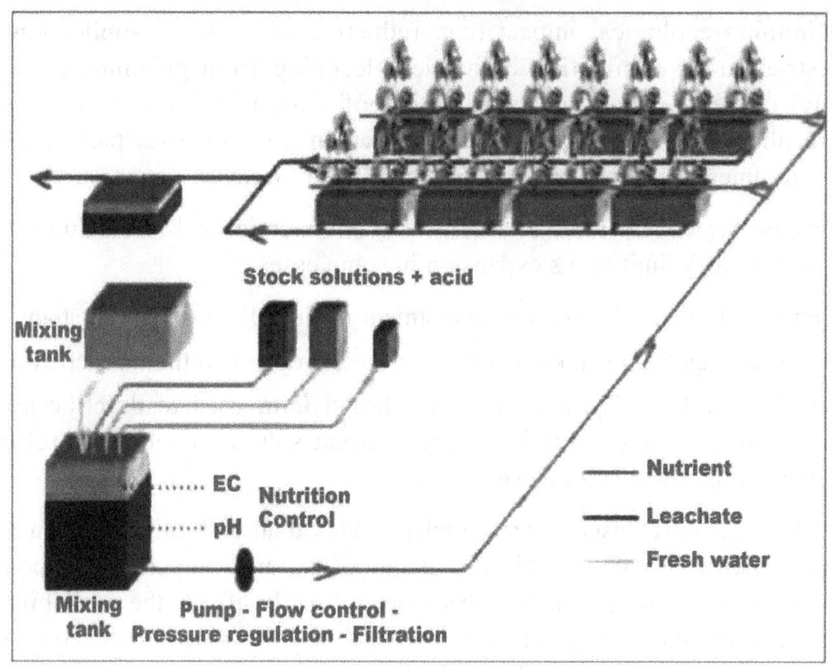

Open loop soil-less culture system

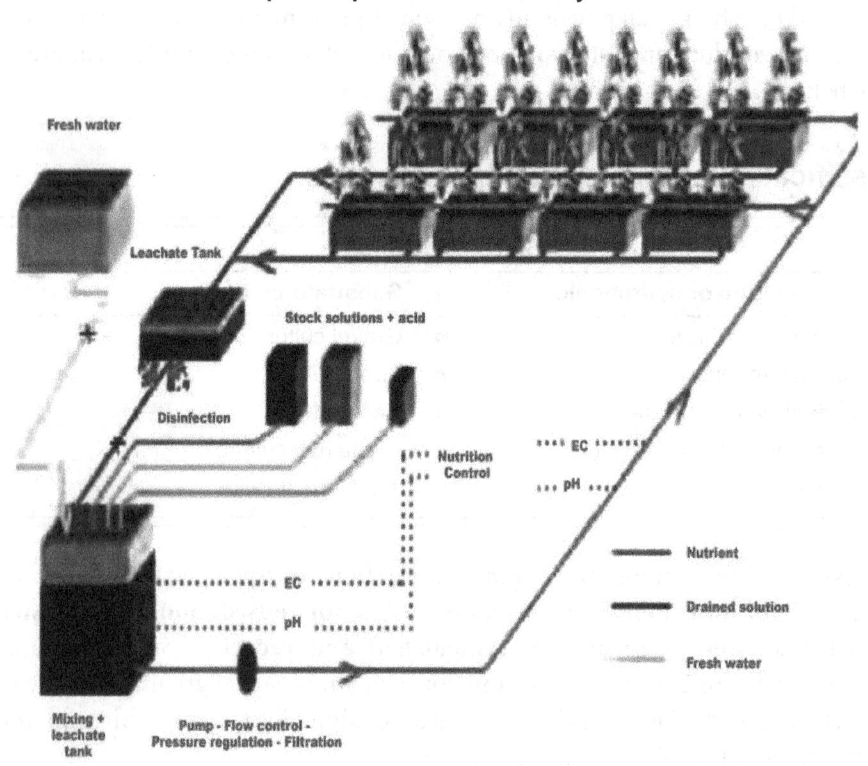

Closed loop soil-less culture system

The following are the functions of the general hydroponic system:

1. It should provide temperature control.
2. Reduce water loss by evaporation.
3. Reduce disease and pest infestation.
4. It should protect crops against weather factors like wind and rain.

Deep water culture (DWC)

Professor W.F. Gericke, University of California proposed and designed DWC in 1929 for the first time for commercial purposes. The system comprises of a bucket filled with nutrient solution, covered with a net and a cloth on which a thin layer of sand (1 cm) is placed to support the plants. Plant roots are suspended in the nutrient solution. Instead, the bucket can be covered with a lid and the plants in net pots are suspended from the centre of the cover. This system sometimes create hypoxic conditions near the root zone due to the limited air water exchange area compared to the volume of the solution and low diffusion of oxygen in the water.

This constraint has now been overcome by pumping air oxygenated air in the nutrient solution or by adopting recirculating deep water culture (RDWC) system which use a nutrient reservoir to supply nutrients to multiple buckets. In RDWC system, the reintroduced water is is broken up in to the reservoir and aerated with the help of spray nozzles.

Float Hydroponics

In this system, plants are raised on trays floating in tanks filled with nutrient solution. Currently, this technique is commercially utilized for the production of leafy greens like lettuce, chicory, lettuce *etc.* and aromatics such as basil, mint *etc.*

The system requires the low set up and management costs and the little automation is needed for monitoring and adjusting the nutrient solution. The use of large volume of nutrient solution helps in buffering the temperature and reducing the frequency of adjustment and restoration of the solution. However, utmost care must be taken to maintain O_2 level in the range of 5 to 6 mg per litre.

Float hydroponics may accommodate a single tank or multiple tank system in greenhouses. Single tank system covers complete growing the area and allows the automation of some operations like placement and removal of floating trays, while multiple tank system include several tanks of more than or equal to 4 m^2 (2 m x 2 m). The use of multiple tanks has an advantage of reducing operational mistakes and managing diseases efficiently.

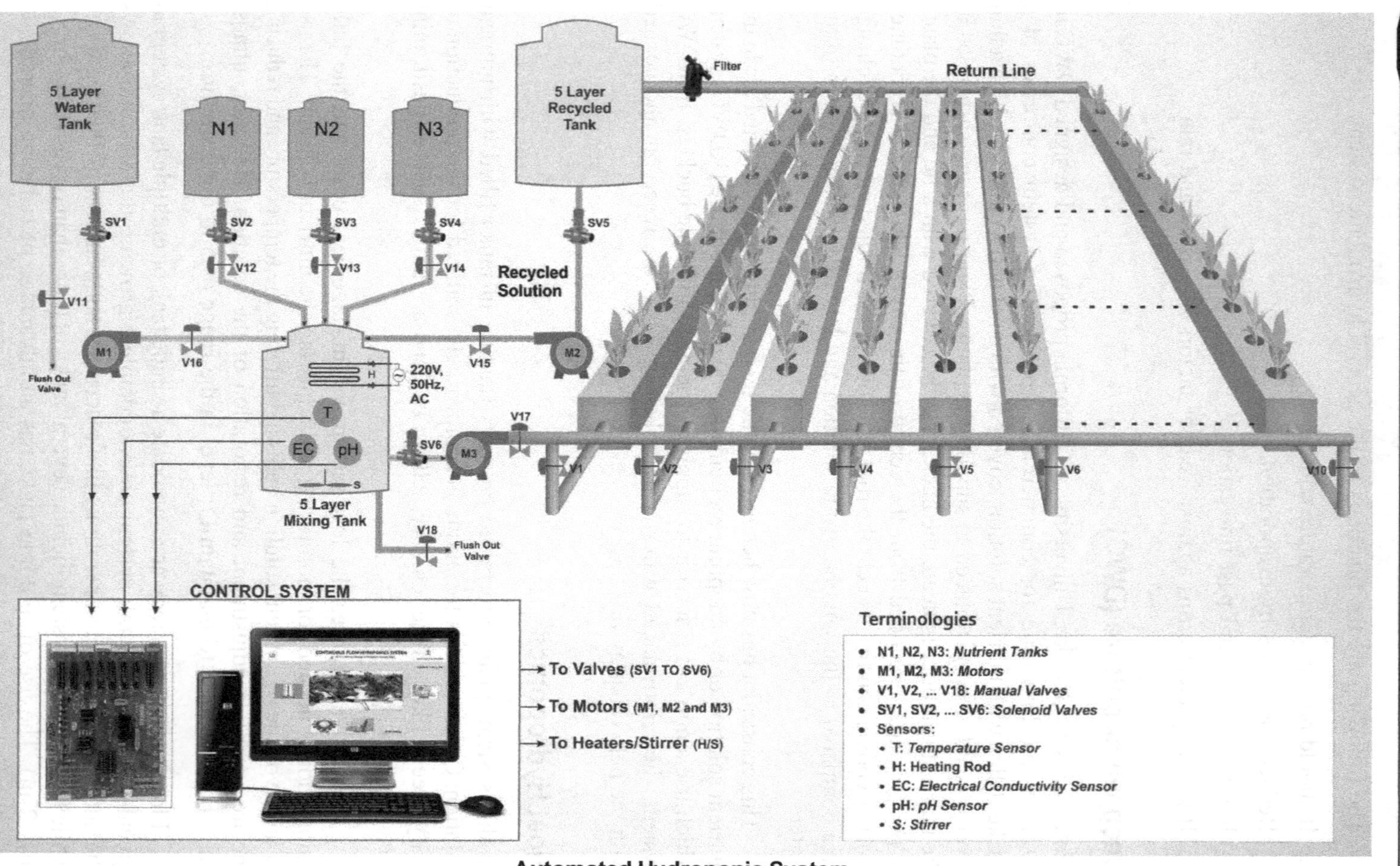

Automated Hydroponic System

Nutrient Film Technique (NFT)

In this system, a very thin layer of nutrient solution is allowed tom flow through water channels (also known as gullies, troughs or gutters), in which the roots of plants are exposed continuously for nutrition gain. The arrangements for the supply of nutrient solution are made from the elevated end of channel so that the solution flows down through the channels keeping the roots completely wet. Floor or benches or racks holding these water channels may be provided with some elevation foe easy movement of solution. The thin water stream of 1 to 2 mm depth is generally maintained to ensure sufficient oxygenation of the roots because of continuous exposure of thick root mat to the air. At the lower end of the channels, the solution is drained to a large catchment pipe, which carries the solution back to the reservoir for recirculation.

Water channels for the system can be made up of plastic material such as polyethylene liner, polyvinylchloride (PVC) and polypropylene with a rectangle or triangle shaped section. Care must be taken to keep the base of the channel flat so as to maintain a shallow stream of nutrient solution. Length of channel should not exceed over 12 to 16 m. Now a modified NFT system *i.e.* Super Nutrient Film Technique (SNFT) has been developed to overcome the problems being faced by the growers. In this system, distribution of nutrient solution is achieved through nozzles arranged along the channel, which ensures adequate supply of nutrients to the growing plants as well as oxygen.

Nutrient delivery in the system could be continuous during 24 hour cycle or mabe be intermittent. In case of intermittent nutrient supply, watering is accomplished alternatively with dry period for improving the oxygen level near to the root system. Continuous recirculation of the nutrient solution is generally maintained during day hours under natural light from dawn to dusk and the flow of nutrients is switched automatically off during night. The crop plants selected for cultivation under NFT system are required to grown first in small pots or plugs or in rockwool cubes and once seedlings attain sizeable root and shoot system, are placed in the channels of system.

NFT has many benefits even than that of other soil-less cultures, as this system don't use any substrate for cultivation and requires less volume of nutrient solution thus ensures better saving of water and nutrients with minimum impact on environmental. In this system, costs related to the disposal of the substrate can also be saved. However, nutrient solution because of low water volume may become prone to temperature changes along the channel and during growing seasons of a crop. Moreover, NFT has very little buffering against interruptions in water and nutrient supplies and considerable risk of the spread of root borne diseases. Technically many crops are suited for cultivation in a NFT system, but it works best for crops with a span of 30-50 days like lettuce, because plants become ready for harvesting prior to the condition where root mass of crops may fill the channel.

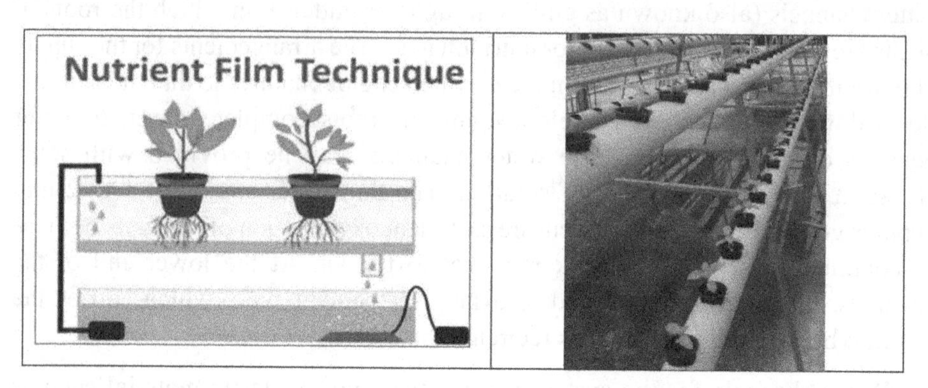

Deep Flow Technique (DFT)

DFT is another system of water culture in which roots of crops are continuously exposed to moving water and nutrients. In this system, nutrient solution remains continuously flowing at a depth of 50-150 mm. the presence of large water volume in the system helps in simplifying the control over nutrient solution and buffering the temperature thus making the system suitable for areas facing problems of temperature fluctuation in the nutrient solution. The width of the channels in a DFT system is usually kept about 1 m and the plants are grown on polystyrene trays which float on the water or rest on the channel sidewalls.

Aeroponics

Aeroponics is a system of growing plants with the root system suspended in a fine mist of nutrient solution applied continuously or intermittently. Plants are placed in holes on polystyrene panels using polyurethane foam. The panels are kept horizontally or on a slope and fixed over a metal frame with close arrangements having square or triangular section.

Water and nutrients are supplied by spraying onto the hanging roots with an atomized nutrient solution with the help of sprayers, misters or foggers inserted in PE or PVC pipes. Spraying of solution is usually done for 30-60 seconds and the frequency may vary as per species cultivated, growth stage, season and time of day. For instance, a crop in temperate region during rapid vegetative growth in summer month may need up to 80 sprayings per day. At each administration, the drainage is collected at the bottom of the unit and recirculated.

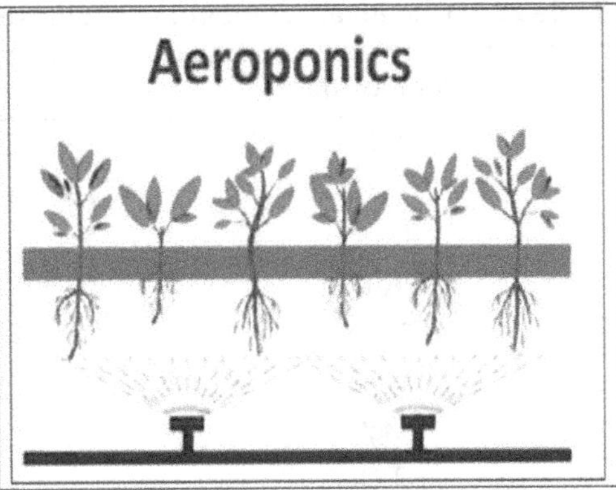

Aeroponics has major advantage of reducing water and fertilizer consumption and simultaneously ensures adequate oxygenation of the roots. However, plants grown under this system may experience severe thermal stresses especially during summer. This system has another disadvantage in terms of its inability to buffer interruptions in the flow of nutrient solution during interruption or failure in the supply of power. Aeroponics can successfully be used be used for small statured crops like lettuce and strawberries and medicinal and aromatic plants. Central Potato Research Institute (CPRI), Shimla has also successfully demonstrated this technology for mini tuber production in potato to meet out the ever increasing demand for seed tubers.

Gravel Culture

Gravel culture is a kind of soil-less substrate culture in which plants are cultivated in pea sized gravels supporting the root system. This type of culture has the advantage of maximizing nutrient delivery as well as aeration to the growing roots. Plants are grown in a gravel bed with the help of water circulation controlled by power head pumps, so it is also known as gravel hydroponics. Gravel is inexpensive, easy to keep clean, drains well and will not become water logged. However, plants needs to be fed continuously with water otherwise chances of plant drying may increase.

Preference should be give to the locally available material in this type of soil-less culture as imported materials may be found costly compared to the local ones. The particle size for local gravel could be 3-7 mm while for the imported gravels the appropriate size is 3-5 mm. It was studies by earlier research that a silicate based material used in the building industry could be utilized in gravel culture, composition of which is Table 28.1

Table 28.1: Chemical composition of silicate based gravels

Composition	Weight (%)
SiO_2	49.10
Al_2O_3	15.60
Na_2O	2.38
K_2O	1.65
Fe_2O_3	2.08
CaO	7.58
MgO	10.42
TiO_2	0.30
FeO	5.04
P_2O_5	0.04
MnO	0.18
Na_2O	2.38
H_2O	4.22
H_3O	1.64
Total	100.23
Specific gravity (SG)	2.76
pH	7.7 (Extraction: 1:2.5 w/w)

Source: Gass, 1960

Similarly, composition of local gravels at a particular location can be studied and used in gravel culture.

Sand Culture

Sand culture is a method of growing plants hydroponically without the use of soil i.e. in sand. It is a variation of gravel culture where sand is primarily used to provide anchorage to plants in the grow bed or tray. Sand used in this type of culture is finer than the gravel culture. Though, sand is cheap and easily available, however it is heavy and does not have good water holding capacity. It is also recommended to sterilize sand prior to its use in the system. In this culture, big vessels are filled with acid and distilled water washed pure sand. Nutrient solution is supplied MIS from time to time as per the requirement of a crop. Plant roots are naturally aerated in this system.

Bag Culture

In bag culture, plants are grown in a soil-less medium contained in lay flat or upright polyethylene bags and fed with liquid fertilizers through drip irrigation lines. Media can be peat/vermiculite, sawdust, rock wool, rice hulls, pine bark, peanut hulls or mixes of any of those. Bag culture is adaptable to the aggregate hydroponics fed with liquid fertilizers as well as standard soil-less culture fertilized with compost based potting mixes. Bag culture is well suited to the production of upright and viny crops like tomato, cucumber, pepper *etc*. One problem with bag culture is the potential for excess fertilizer solution to leach out of the drainage holes in the bags and into the greenhouse soil.

Container Culture

Container culture refers to the practice of growing plants exclusively in containers filled with soil-less medium. A container is a small, enclosed and usually portable object and very good option for displaying live plants particularly those ones with aesthetic values like ornamentals. Containers range from simple plastic pots (pot, box, tub basket, tin, barrel, hanging basket *etc.*) to complex automatic watering systems. This system offers flexibility in design and hence very popular among growers as well as customers. Container culture could well be adopted on porches, front steps and on rooftops. Sub-irrigated planters (SIP) are a type of container that may be used in container culture Potting mix or media to be used in this system should be loose and have good drainage capacity as well as ability to offer proper aeration for roots.

Trough Culture

Trough culture refers to cultivation of plants in a soilless mixture (e.g. coconut fiber, clay pellets, vermiculite, perlite, or rockwool) in raised troughs. The trough system also comprises of a solution storage tank for storing nutrient solution, a solution feeding pipe and a solution return pipe, which serve as communicators between solution storage tank and the cultivation trough. In this system, one or more cultivation furrows can be made on one side or both sides of the cultivation trough.

29

GOOD AGRICULTURAL PRACTICES (GAP) FOR PROTECTED CULTIVATION

Good Agricultural Practices (GAP) are the practices that address environmental, economic and social sustainability for on-farm processes and result in safe and quality food and non-food agricultural products.

GAP primarily involve the application of good management practices to maintain the consumer confidence in food quality and food safety by taking into account the optimal use of inputs to ensure worker health and minimizing detrimental environmental impacts on farming operations.

The objectives of GAP codes, standards and regulations are:

- To ensure safety and quality of produce in the food chain.
- To capture new market advantages by modifying supply chain governance.
- To improve the use of natural resources, workers' health and working environment.
- To create new market opportunities for farmers and exporters.

Four Pillars of GAP

1. Economic viability
2. Environmental sustainability
3. Social acceptability
4. Food quality and safety

Major Principles of GAP

1. Traceability
2. Record keeping and self inspection

3. Varieties and rootstocks
4. History of site and its management
5. Soil and substrate management
6. Fertilizer use
7. Irrigation/Fertigation
8. Crop protection
9. Harvesting
10. Produce handling
11. Waste & pollution management, recycling and re-use
12. Worker health, safety and welfare
13. Environment issues
14. Complaint form

Key Elements of GAP

1. Prevention of problems before they occur
2. Risk assessments
3. Commitment to food safety at all levels
4. Communication throughout the production chain
5. Mandatory employee education program at the operational level
6. Field and equipment sanitation
7. Integrated pest management
8. Oversight and enforcement
9. Verification through independent, third-party audits

India – IndiaGAP

The Agricultural and Processed Food Products Export Development Authority of India had initiated the development of an IndiaGAP standard with objective to gain benchmarked recognition with GLOBALGAP so as to open the European market to Indian agricultural producers.

In India, Bureau of Indian Standards (BIS) has taken initiatives to develop its own standards to be followed by institutions and companies, etc. The draft Indian Standard Good Agricultural Practices – IndiaGAP (Part 1- Crop Base) takes into account not only the quality and quantity of the crop obtained from a unit area but also the care and attention gone into integrating pre-harvest practices like soil & water management, nutrient management and pest management, harvesting, post harvest handling and other logistics. The objective is to ensure food safety, occupational health/safety/welfare, and wherever possible, animal welfare. The entire operation is intended to make farming practices environment friendly. For the purpose of verification, a graded pattern given below (**Table 29.1**) shall be followed for grant of IndiaGAP licence:

Table 29.1: Proposed Graded Pattern for verification under BIS IndiaGAP

Category of Licence	Major compliances	Major compliances
IndiaGAP – A	100 %	90 %
IndiaGAP – B	100 %	80 %
IndiaGAP – C	100 %	75%

BIS India GAP certification shall be as prescribed under the provisions of Bureau of Indian Standards Act, 1986 and Rules and Regulations framed there under. The details of the conditions under which the licence may be granted to producer (individual grower and/or member of a grower group) may be obtained from the Bureau of Indian Standards. Food and Agriculture Department of Bureau of Indian Standards has formulated various standards since inception and has so far developed around 1800 standards in the following areas including – pesticides, sugar, apiary, tobacco, livestock feeds, equipment, stimulant foods, soil quality and fertilizers, food additives, spices and condiments, processed fruit and vegetable, agricultural tractors, fish and fisheries products, oil and oilseeds, drinks and carbonated beverages, food hygiene, safety management food grains, starches and ready to eat foods, irrigation systems, farm implements, slaughter house and meat, dairy products and equipments, agriculture and food processing equipments, agriculture management and systems, biotechnology and specialized products. These standards are for products, methods of test, code of practice, terminology, symbols and systems. However, the BIS – IndiaGAP draft document still needs to be finalized before it could he used as standard.

Potential Benefits of GAP

- To improve the safety and quality of food and other agricultural products by adopting and monitoring GAP appropriately.
- It can potentially lessen the risk of non-compliance with national and international regulations, standards and guidelines [Codex Alimentarius Commission (CAC), World Organization for Animal Health (OIE) and the International Plant Protection Convention (IPPC) with respect to permitted limits of pesticides, maximum permissible levels of contaminants e.g. pesticides, veterinary drugs, radionuclide and mycotoxins) in food and non-food agricultural

products as well as other chemical, microbiological and physical contamination hazards.

- To promote sustainable development in agriculture and contribute towards meeting national and international ecological and social development objectives.

Challenges Related to GAP:

1. GAP implementation especially record keeping and certification may increase production costs. In this respect, lack of harmonization between existing GAP-related schemes and availability of affordable certification systems has often led to increased confusion and certification costs for farmers and exporters.
2. Standards of GAP can be used to serve competing interests of specific stakeholders in agri-food supply chains by modifying supplier-buyer relations.
3. Small scale farmers need to be adequately informed, technically prepared and organized by governments and public agencies to make them capable enough to grab export market opportunities with playing a facilitating role.
4. Compliance with GAP standards does not always guarantee sustainability of environmentally and socially.
5. Awareness among the growers is very important bring improvement in yield and production efficiencies as well as environment and health and safety of workers. One can also think of such an approach like Integrated Production and Pest Management (IPPM) to safeguard the interest of people as well as environmental safety.

GLOBALG.A.P. - Putting Food Safety and Sustainability on the Map

G.A.P. stands for **Good Agricultural Practice** – and GLOBALG.A.P. is the **worldwide standard** that assures it.

GlobalGAP Certification

GLOBALGAP (formerly known as EurepGAP) has established itself as a key reference for Good Agricultural Practices (G.A.P) in the global market-place. The GLOBALGAP standard is primarily designed to reassure consumers about how food is produced on the farm by minimizing detrimental environmental impacts of farming operations, reducing the use of chemical inputs and ensuring a responsible approach to worker health and safety as well as animal welfare.

Primary producers, retailers like Sainsbury and Marks & Spencer, grower-associations, traders and their organizations can be certified against GLOBALGAP standards. Sectors such as fruits, vegetables, potatoes, salads, cut flowers, nursery stock, etc. are covered under this standard.

Purpose

GLOBALG.A.P. members create private sector incentives for agricultural producers worldwide to adopt safe and sustainable practices to make this world a better place to live in for our children.

Mission

Globally connecting farmers and brand owners in the production and marketing of safe food to provide reassurance for consumers. We lay the foundation for the protection of scarce resources by the implementation of Good Agricultural Practices with a promise for a sustainable future.

Benefits of GLOBALGAP

- Ensures retailer, consumer, employee confidence through responsible and sustainable production and proactive commitment to food quality.
- Supports the basic principles of HACCP- Hazard Analysis and Critical Control Points (BRC- British Retail Consortium).

Reassuring consumers around the world that the food products they are buying are safe and responsibly produced – in terms of their environment impact as well as the health, safety and welfare of workers and animals – has reached levels of unprecedented importance in today's society. Because validating such efforts on an individual, business-by-business basis is an extremely complex undertaking, third-party certification schemes are available. One of the most internationally recognized initiatives in this respect is Global GAP, a non-governmental organization (NGO) that sets voluntary standards for the certification of agricultural products all across the globe.

GAP for Protected Cultivation

Intensive cultivation of crops as followed under protected culture requires some set of GAP protocols as protected cultivation is generally involved in excessive use of chemicals since the stakes are high due to intensive inputs and high expectations on quality front.

Site Selection

- Site should be selected keeping in view of production costs and quality of produce as these production factors are majorly governed by environment.
- Transportation cost of greenhouse produce should also be taken into consideration while selection site.

Climate Conditions

- Solar radiation and other climatic parameters particularly air temperature should ve available all he year round.
- Tropical and subtropical areas vulnerable to strong winds should appropriately be protected by planting wind breaks.
- Provision s must be ensured for protection against heavy rains and high solar radiation.
- Cladding materials should also be selected appropriately depending upon the climatic conditions of the region.

Insect-proof Nets/screens

- Accurate size of insect-proof nets should be used to avoid any kind problem for ventilation in the structure.
- Insect proof nets with smaller thread diameter should be preferred for more porosity leading to better ventilation.
- Photo-selective nets may be selected as they provide additional protection against insect-pests.

Ventilation

- There must be provision for vent at the ridge, on the sidewalls and the gable particularly in naturally ventilated polyhouses (NVPH).
- The area equivalent to 15-30% of floor area of protected structure is generally recommended for effective ventilation.

- If regions are not prone to high wind speed, the mechanism of natural ventilation may be adopted to create congenial environment for crops inside the structures.
- Depending upon the type of structures and requirement for ventilation, the criteria of getting highest ventilation rates per unit ventilator area should be considered suitably. For instance, when flap of roof ventilators faces the wind ventilation to a level of 1000% can be achieved, which is followed by flap ventilators facing away from the wind ensuring 67% ventilation while the lowest rates are achieved with rolling ventilators (28%).
- Exhaust fans and blowers can supply high air exchange rates whenever needed.

Evaporative Cooling

The phenomenon of evaporative cooling helps in lowering temperature and vapour pressure deficit hence, plays important role to achieve desirably lower temperature inside protected structure compared to the outside air temperature. This helps in maintaining uniform growing conditions throughout the greenhouse.

GAP Recommendations 1 for Fogger System

- The pressure of fogger system can be as high as 40 bars or 50.99 kg/cm^2 or low as 5 bars or 5.10 kg/cm^2 depending upon the requirement for greenhouse.
- The nozzles of the fog system should be installed at the appropriate height to avoid droplets falling on crop canopy or ground.
- The maximum aperture of vent *i.e.* 20% should be kept open while operating fogger system.
- Nozzles fitted with fans generally provide 1.5 times better evaporation ratio and three times wider cooling area than nozzles without fans.

GAP Recommendations 2 for Fan and Pad

- The general recommendation for pad thickness is 200 mm and the material used in cooling pad should have a high surface, good wetting properties and high cooling efficiency.
- The distance between fan and pad should not exceed 40m and the pad area should be approximately 1 m^2 per 20-30 m^2 of greenhouse area.
- Fans should be placed on the leeward side of the greenhouse and the distance between fans should not exceed 7.5-10 m. If fans are installed on the windward side, then an increase of 10% in the ventilation rate is compulsory.

- The water flow through the pad should be turned on first before starting the cooling system and the fan should be turned off while stopping the cooling system to check the pads from clogging.
- An airflow at the rate of 120-150 m^3/m^2 greenhouse area per has been found satisfactory for effective working of fan and pad cooling.

Dehumidification

- Side curtains of the greenhouse should be opened to allow the excess moisture to escape through ventilation.
- Ensure to remove any excess sources of water from inside the greenhouse structure.
- Thermal screens can be used to prevent radiative heat loss from crop surfaces during night hours.
- Exhaust fans can be kept operative to improve air circulation.

Climate Control

- Maintenance of devices, pumps, valves, ventilators *etc.* prior to each cropping period is essential for effective working of the system.
- Thermostats/sensors placed inside the structure must be protected from direct exposure to sunlight.
- Solar energy should appropriately be trapped during winter season by delayed opening of ventilation.
- Conduct monitoring of environmental control system or thermostats at regular intervals.
- If thermal screen is in use, try to ensure the opening of screen first than the vents) for reducing humidity.
- Apply CO_2 at least to ambient concentration (*i.e.* 340–370 ìmol mol^{-1}) if available. It does not reduce energy use but significantly contributes to crop growth and production.

Energy Efficiency

- Energy-efficient models of protected structures should be selected.
- Carry out maintenance of hardware components of greenhouse at regular intervals.
- Doors of the structure must be kept closed and make sure to seal air leakages, replace broken cladding material and ripped screens.

- Cladding materials low infra red (IR) transmission should be selected for the structure.
- In regions with low average or low night temperatures, moveable thermal screens should be used.
- Greenhouses should be fabricated with large vent openings in windward direction openings particularly in NVPH.
- For cooling purpose, use fogging system or pad and fan cooling in combination with shading nets.

GAP Practices for Water Management

- Water bodies should be maintained to prevent surface run-off from contaminating water supply.
- Run-offs should be managed properly and the desirable run-off water should be stored properly for increasing water use efficiency.
- Cleaning of water reservoirs must be ensured at least once in a year.
- Adopt water management plan to optimize water usage and reduce wastage of water by maintaining irrigation equipment properly for high efficient working.
- If possible, use minimum energy utilization techniques like zero energy drip irrigation system.

Improving Irrigation Efficiency

- It is mandatory to have an idea of the soil moisture in root zone prior to scheduled irrigation, which is to be maintained at field capacity.
- Check the irrigation system regularly for any type of clogging.
- Use pressure compensating drippers in the irrigation system.
- Pay regular attention to Delivery efficiency (DE), Farm efficiency (FE) and Watering efficiency (WE) or distribution efficiency. Where, DE is the ratio between the volume of water delivered to the farm and the volume of water taken from the source; FE is the ratio between the water volume distributed by the farm on irrigated land and the volume of water delivered to the farm or taken directly from a water source on the farm; WE the ratio between the water volume retained in the soil layer and usable by plants and the watering volume.

Soil Fertility and Plant Nutrition

- Pay utmost attention to the synchronization of fertilization to actual mineral uptake by the plants for optimal crop growth.

- Make sure to enrich soil media with organic matter.
- Fertigation to a crop should be scheduled as per the crop requirements at different stages of growth and soil nutrients status.

GAPs for Harvesting

- Harvesting of the produce must be carried out at appropriate maturity stage so as to retain optimum eating quality at consumer end.
- Early harvesting should be avoided as it may result to decay, water loss and/or lack of sugars. Similarly, late harvesting may cause premature senescence, susceptibility to decay and over-ripe product with decreasing taste, flavour and texture.
- Prioritize the optimum stage of harvest depending on the location and on the distance and transport time from the production site.
- Avoid physical damage to produce during harvesting and transport to the packing facility, as any physical damage in the form of cuts, bruises and scarring may lead to infection.
- Tools used for harvesting must be clean, sharp and sanitized regularly.

GAPs for Post-harvest Management

- Profitability
- Consumer
- Supply chain
- Technology
- Pre-harvest
- Handling
- Optimal product temperature
- Storage
- Cleanliness and sanitation
- Sorting and grading
- Packaging
- Internationally accepted pallets
- Product knowledge
- Quality assurance systems
- Staff education and training

30

COST CONCEPT AND ECONOMIC ANALYSIS OF PROTECTED CULTIVATION

With the advent of modern technologies, the scenario of protected cultivation industry in India is changing at a fast rate. Now, it is not only the question of providing enough produce, but also to ensure quality production throughout the year that are acceptable and competitive in international market. But due to erratic behaviour of weather, the crops grown in open field are often exposed to fluctuating levels of temperature, humidity, wind flow *etc.*, which ultimately affect the crop productivity and quality adversely. Protected cultivation being the most efficient means to overcome climatic diversity, has the potential of fulfilling the requirements of small growers as it can increase the yield manifolds and at the same time improve the quality of the produce significantly as per the demand of the market. In the recent times, introduction of parthenocarpic varieties in cucumber for instance has revolutionized its cultivation under protected culture in India. Simultaneously, implementation of protected cultivation through various assistance under Mission for Integrated Development of Horticulture (MIDH), *Rashtriya Krishi Vikas Yojna* (RKVY), National Horticulture Board (NHB) and many more at state level have bolstered the adoption of protected cultivation across the country.

Economics of a greenhouse is governed by various facilities as well the location selected for greenhouse structures. Another important economic component is structural dimensions of a greenhouse with the facility of partially or fully controlled environment enabling round the year and off-season production leading to higher net realization. Greenhouse technology is emerging as an industry and ensuring self-employment as well as employment to labourers because of the possibility of producing crops round the year. Different types of structure have differential response towards production potential depending upon crops, which ultimately decide the amount of revenue generated from a particular protected structure.

Points to Ponder for Improving the Economy of Protected Cultivation

Selection of appropriate location with good accessibility to transportation facilities, when combined with the following factors can direct or indirect investment thereby minimizing the cost.

1. Availability of plenty of sunshine hours.
2. Presence of mild winters.
3. Stability in weather parameters over a period of time.
4. Appropriate level of humidity in the environment.
5. Availability of good quality water with appropriate level of EC and pH, which will obviate expenditure on installation of RO plant.

In addition to selection of site, there are also some other factors which can be taken into consideration for improving the economy through protected cultivation and these are:

1. Mechanization of certain horticultural operations training, pruning, harvesting, grading, packaging etc.
2. Efficient use of labour.
3. Proper packing and marketing of produce i.e. branding of produce.
4. Managerial Skills.

Cropping pattern does have remarkable impact on the economic gain from protected cultivation. For instance, selection of crops for the year round production under protected environment particularly in naturally ventilated polyhouses is very important to get better returns out of cultivation. while, hi-tech greenhouses provide an added benefit over low tech greenhouses in relation to cultivation of crops as per the active demand of consumers not only in the nearby markets but also at distant markets. Thus, demand driven production of a particular crop will definitely ensure higher returns. Certain unprecedented situation (s) as observed during corona pandemic (COVID 19) may also offer opportunities to the growers for thinking of alternative crops particularly the ones becoming popular amongst consumers at certain point of time. Simultaneously, there has also been an increase of demand for protective food, which is actually allowing the growers to change the plan of action involving cultivation of vegetables and fruits.

It becomes imperative to work out the economic feasibility of greenhouse production system taking all the factors into account so that an optimal tool for decision support in real life situation could be worked out in helping out the decision farmers and makers.

Cost Concepts and Components of Cost

Broadly, the costs are divided into 2 broad categories *viz.*, Fixed costs and Variable costs.

Fixed costs: These are fixed i.e. Fixed cost remain the same irrespective of level of production. These costs remain invariant in the short run but in the long run there are no fixed costs as all the inputs may vary. Fixed cost in soil based greenhouse production system includes greenhouse structure, red soil, plant support system, interest on fixed capital and rental value of land. These are also known as indirect costs, sunk costs and overhead costs.

The actual values on fixed investment are subjected to amortized accounting by adopting certain assumptions and these are:

Assumptions for the calculation of fixed components of cost (Amortized Cost)

Particulars	Useful life (yrs)	Remark
Polyhouse Structure	10	*Conditional life of red soil has been considered equivalent to that of structure's life assuming that sufficient organic matter will be incorporated into it over the period of time.
Red soil*	10	
Plant support system	5	

Fixed Cost = Rental value of land + Interest on fixed capital + Amortized Cost

Variable costs: The costs that vary with the production or the costs which vary with the level of output. These include costs incurred on seeds, fertilizers, plant protection chemicals, micronutrient, bio-fertilizers and labours etc. These are also known as operating costs, working cost, direct costs, prime costs, circulating costs and running costs.

The cost of production of a crop is considered at three different levels *viz.*, Cost A, Cost-B and Cost-C. These cost concepts are generally followed in the studies of production cost of crops.

The input items included under each category of cost are given below.

Cost-A: Actual paid-out costs for owner cultivator, inclusive of both cash and kind expenditure which include following cost items,

1. Hired human labour : a) Male b) Female
2. Total bullock labour a) Owned b) Hired
3. Seeds

4. Manures
5. Fertilizers
6. Insecticides and pesticides
7. Irrigation charges
8. Land revenue and other taxes
9. Depreciation or capital assets
10. Transport
11. Interest on working capital

In broad sense;

Cost A = Variable cost + Interest on working capital [7% Interest on working capital (Variable Cost) per annum]

Based on crop duration, the actual interest on working capital can be worked out.

Cost B: If the amount invested in purchase of land would have been put in some other long term enterprise or in a bank, it would have yielded some returns or interest. But due to the investment of the amount in purchase of land, the farmer has to scarify returns or interest that he would have otherwise gained. As such this loss is considered as interest on fixed capital. Similarly, the hypothetical interest that the capital invested in farm business would have earned, if invested alternatively is also considered as cost. Rental value of land and interest on fixed capital represent imputed costs which are added to Cost A to give Cost B

Cost B = Cost A + imputed rental value of land (1/15th of Yield Appraisal/ Gross Returns) + Interest on fixed capital (10% of fixed cost) + Amortized Cost.

Cost C: It is the total cost of production which includes all cost items including actual as well as imputed. The value of holding's own labour is to be imputed and added to cost B to workout Cost C.

Cost C = Cost B + imputed value of family human labour.

Gross Returns = Production x selling rate

Net Returns = Gross Returns – Cost C

Benefit Cost Ratio (BCR) = Net Returns/ Cost C

Summary of cost analysis for protected cultivation:

Production from greenhouse system (kg)	Selling Rate (Rs.)	Gross Returns (Rs.) [1*2]	Amortized Cost for a single season	Cost A (Variable cost + Interest on working capital)	Cost B (Cost A + imputed rental value of land + Interest on fixed capital + Amortized Cost)	Cost C (Cost B + imputed value of family human labour)	Net Return (Rs.)	BCR (Net Return/ Cost C)
1	*2*	*3*	*4*	*5*	*6*	*7*	*8*	*9*

Government policies also influence the financial returns from the crops. The component of protected cultivation is being strengthened under **Mission for Integrated Development of Horticulture (MIDH)** by Government of India through 50% subsidy to the farmers. Incentives in terms of subsidy to the tune of 65 and 75% are further disseminated by Government of Gujarat State (India) to encourage the farmers for adopting protected cultivation by adding its share of 15 and 25% in Union Government subsidy depending upon socio-economic status of the farmers.

Cost norms and pattern of assistance under Mission for Integrated Development of Horticulture (MIDH) during XII FY Plan for Protected Cultivation

Sr. No.	Item	Cost Norms (Rs.)	Pattern of Assistance
Green House structure			
1.	**Fan & Pad system**	1650/m^2 (up to area 500 m^2) 1465/ m^2 (>500 m^2 up to 1008 m^2) 1420/ m^2(>1008 m^2 up to 2080 m^2) 1400/ m^2 (>2080 m^2 upto 4000 m^2)	50% of cost for a maximum area of 4000 m^2 per beneficiary.
	*** Above rates will be 15% higher for hilly areas.**		
2.	**Naturally ventilated system**		
	i) Tubular structure	1060/ m^2 (up to area 500 m^2) 935/ m^2(>500 m^{2}upto 1008 m^2) 935/ m^2 (>500 m^2 up to 1008 m^2) 890/ m^2 (>1008 m^2 upto 2080 m^2) 844/m^2 (>2080 m^2 upto 4000 m^2)	50% of cost limited 4000 sq. m. per beneficiary.
	*** Above rates will be 15% higher for hilly areas.**		
	ii) Wooden structure	540/ m^2 621/ m^2 for hilly areas	50% of the cost limited to 20 units per beneficiary (each unit not to exceed 200 m^2).
	iii) Bamboo structure	450/ m^2 518/ m^2 for hilly areas	50% of the cost limited to 20 units per beneficiary (each unit should not exceed 200 m^2).

[Table Contd.

Contd. Table]

Sr. No.	Item	Cost Norms (Rs.)	Pattern of Assistance
3.	**Shade Net House**		
	(a) Tubular structure	710/ m^2 816/ m^2 for hilly areas	50% of cost limited to 4000 m^2 per beneficiary.
	(b) Wooden structure	492/ $m^2$566/ m^2 for hilly areas	50% of cost limited to 20 units per beneficiary (each unit not to exceed 200 m^2).
	(c) Bamboo structure	360/ m^2 414/ m^2 for hilly areas	50% of cost limited to 20 units per beneficiary (each unit not to exceed 200 m^2).
4.	Plastic Tunnels	60/ m^2 75/ m^2 for hilly areas.	50% of cost limited 1000 m^2 per beneficiary.
5.	Walk in tunnels	600/m^2	50% of the cost limited to 5 units per beneficiary (each unit not to exceed 800 m^2).
6.	Cost of planting material & cultivation of high value vegetables grown in poly house	140/ m^2	50% of cost limited to 4000 m^2 per beneficiary.
7.	Cost of planting material & cultivation of Orchid & Anthurium under poly house/ shade net house.	700/ m^2	50% of cost limited to 4000 m^2 per beneficiary.
8.	Cost of planting material & cultivation of Carnation & Gerbera under poly house/ shade net house.	610/ m^2	50% of cost limited to 4000 m^2 per beneficiary.
9.	Cost of planting material & cultivation of Rose and lilium under poly house/ shade net house	426/ m^2	50% of cost limited to 4000 m^2 per beneficiary

Solved Numerical

Calculation of net returns and Benefit-Cost Ratio for greenhouse cucumber raised under naturally ventilated polyhouse- tubular structure of 1000 m^2 area using following information;

Crop Span of greenhouse crop: 100 days & Production per 1000 m^2= 14 tonne

Per unit establishment cost of polyhouse

			Rs. per 1000 m^2
Particulars	**Without subsidy**	**With 65% subsidy**	**With 75% subsidy**
Polyhouse structure @ 935 per m^2	935000	327250	233750
Interest on fixed capital (10% of Fixed cost)	93500	32725	23375

Amortized Cost:

S. N.	Fixed Investment	Rate (Rs.)	Requirement	Total Cost	Useful Life (yrs)	Season specific Investment (Rs.)
1.	Naturally ventilated system- Tubular Structure	935/ m^2	1000	935000	10	25713
2.	Red soil 70 tractor trolleys or 9000 cubic ft	800/100 cubic feet	90	72000	10	1980
3.	Plant support system @ 10/m^2	10/m^2	1000	10000	5	550
	Total					**28243**

Calculation of components of variable cost:

Quantity of Inputs and their respective cost:

Fertilizers	**Requirement per 1000 m^2 (kg)**	**Rate (Rs.)**	**Cost (Rs.)**
19:19:19	10.0	88.00	880.00
00:52:34	2.20	111.20	244.64
00:00:50	1.44	70.00	100.80
$CaNO_3$	8.00	42.00	336.00

[Table Contd.

Contd. Table]

Fertilizers	Requirement per 1000 m^2 (kg)	Rate (Rs.)	Cost (Rs.)
12:61:00	3.76	92.00	345.92
Urea	1.20	6.30	7.56
00:00:50	6.40	70.00	448.00
Urea	2.40	6.30	15.12
$CaNO_3$	7.20	42.00	302.40
00:52:34	4.20	111.20	467.04
00:00:50	3.60	70.00	252.00
Total (I)			**3399.00**

Particulars	Rate (Rs.)	Requirement	Total Amount (Rs.)
Labour (1 labour is sufficient to carry out the operations in 1000 m^2 NVPH)	178	100 man days	17800
Pesticides (For 1000 m^2)	2500	1	2500
Vermicompost 0.4t (8 Bags)	200	8	1600
Requirement of Formaldehyde (litres)	50	75	3750
Application of formaldehyde (For 1000 m^2)	2000	1	2000
Trichoderma viridi (kg)	120	0.5	60
Pseudomonas inflorescens (kg)	163	0.5	82
Micro-nutrients (kg)	200	5	1000
Bed preparation (man days)	178	15	2670
Miscellaneous (For 1000 m^2)	4000	1	4000
Seed Cost (Nos.)	5	3000	15000
Total (II)			**50462**

Calculation of packing material cost:

Production (kg/1000 m^2)	No. of bags of 20 kg capacity (Packing of 20 kg is made for local market)	1 kg accom-modates no. of bags	No. of bags required	Cost of 20 kg packing	Cost of packing material
1	2	3	4 (2/3)	5	6 (4 x 5)
14000.00	700	35	20	160	**3200 (Total III)**

Calculation of UV stabilized supporting string

UV stabilized supporting string (Nos.)	Wt of single string (g)	Plants per 1000	Wt of string (kg)	Cost of string per kg	Total Cost of string per 1000	Life of string (Season)	So, Cost Per Season (Rs.)
1	1.38	2900	4.0	220	880	3	**293 (Total IV)**

Hence, **Total Variable Cost** = Total (I + II + III + IV)

= 3400 + 50462 + 3200 +293

= **57355**

Summary of economic analysis for greenhouse cucumber production system:

Production from green house system (kg)	Selling Rate (Rs.)	Gross Returns (Rs.)	Amortized Cost for a single season	Cost A (Variable cost + Interest on working capital @ 7%)	Cost B (Cost A + imputed rental value of land + Interest on fixed capital + Amortized Cost)	Cost C (Cost B + imputed value of family human labour)	Net Returns (Rs.) [3-7]	BCR (Gross Returns/ Cost C) [3/7]
1	2	3	4	5	6	7	8	9
14000	25	350000	28243	61370 (57355 + 4015)	201780 (61370 + 18667 + 93500 + 28243)	201780*	148220	0.73

* In current calculation, family labour has not been taken into consideration

31

UNDERSTANDING THE CONCEPT OF VEGETABLE GRAFTING

Vegetables are nutritionally rich, high-valued crops and remunerative enough to replace subsistence farming. However, they are highly sensitive to climatic vagaries and sudden irregularities in weather factors at any phase of crop growth can affect the normal growth, flowering, fruit development and subsequently the yield. Grafting in vegetable has emerged as a promising surgical alternative over relatively slow conventional breeding methods aimed at increasing tolerance to biotic and abiotic stresses. It provides an opportunity to transfer some genetic variations of specific traits of rootstocks to influence the phenotype of scion. Thus, genetic potential of various rootstocks in vegetable crops has proven to be a better alternative to chemical sterilants against many soil-borne diseases.

What is Grafting?

Grafting is an art and technique in which two living parts of different plants *i.e.*, rootstock and scion are joined together in such a manner that they would unite together and subsequently grow into a composite plant. There are different methods of grafting which can be employed in vegetable crops namely Tongue approach grafting, Hole insertion grafting, Splice grafting, Cleft grafting, Tube grafting.

Purposes of Grafting

1. Imparting disease and pest resistance
2. Avoiding nematode infestation
3. Minimizing the auto toxic effect
4. Providing cold and heat hardiness
5. Improving quality traits
6. Manipulating the harvesting period

7. Reduced fertilizer and agrochemical application
8. Increase yield

For successful rising of the crop under biotic and abiotic stress, precocity, improvement of quality and other horticultural attributes. Further crop wise objective is given in Table 13.1.

Table 31.1: Cropwise objectives of grafting

Crop	Objective (s)
Bitter gourd	Tolerance to Fusarium (*Fusarium oxysporum* f. sp. *momordicae*)
Cucumber	Tolerance to Fusarium wilt, *Phytophthora melonis*, cold hardiness, favourable sex ratio, bloomless fruit
Brinjal	Tolerance to bacterial wilt, (*Pseudomonas solanacearum*) *Verticillium alboatrum, Fusarium oxysporum,* low temperature, nematodes, induced vigour and enhanced yield.
Muskmelon	Tolerance to Fusarium wilt (*Fusarium oxysporum*), wilting due to physiological disorder, *Phytophthora* disease, cold hardiness, enhanced growth
Tomato	Tolerance to corky root (*Pyrenochaeta lycopersici*),*Fusarium oxysporum* f.sp. *radicislycopersici*, better colour and greater lycopene content, tolerance to nematode.
Watermelon	Tolerance to Fusarium wilt (*Fusarium oxysporum*), wilting due to physiological disorder, cold hardiness and drought tolerance

Basic Prerequisites of Grafting

- Root stock
- Scion
- Compatibility
- Grafting aids like grafting clips, tubes, pins, blade, cutter *etc*.

Grafting clips

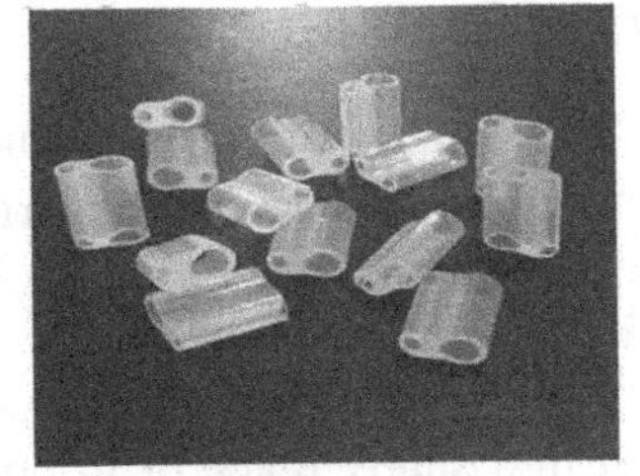
Grafting tubes

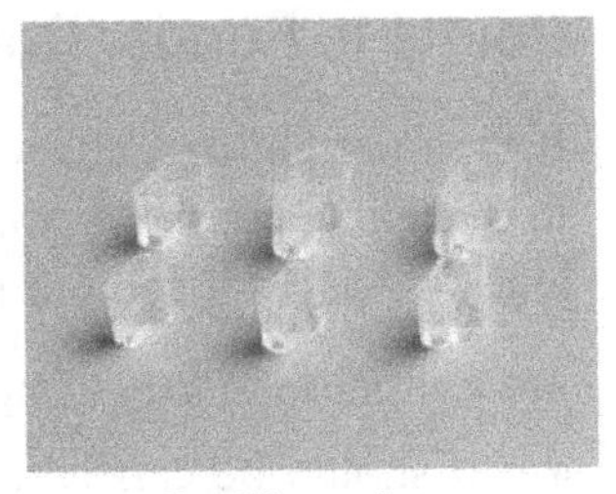
Grafting pins

- **Screen house:** It is used for growing seedlings prior to grafting. There must be provision for double door system at the entrance of screen house. The cladding material for covering the screen house should be made of UV stabilized polyethylene film to prevent UV light penetration.
- **Healing chamber/Grafting chamber:** The healing chamber is a covered structure with controlled humidity and low light. The purpose of healing chamber is to provide congenial temperature and humidity for better union of the grafts. Grafted plants can also be healed under in plastic tunnels maintaining almost near to optimum conditions for better healing. In general, optimum rage of temperature and relative humidity is 25-30 ^{0}C and 85-90%, respectively with low light intensity.
- **Acclimatization chamber:** It is used for hardening of grafts prior to transplanting. This chamber helps to prevent leaf burning and wilting of the just healed seedlings. Grafted seedlings take 7 to 10 days for acclimatization.

Screen House

Healing Chamber

Acclimatization chamber

Healing Process

Union formation

Grafts ready for planting

Points to Ponder During Grafting

It is important to increase the chances of vascular bundle of scion and rootstock to come into contact by maximizing the area of cut surface that are spliced together and pressing spliced cut surface together.

- Cut surface should not be allowed to dry out.
- It should be carried out in a shady place or in polycarbonate house.
- Expose the scion and rootstock to sunlight for 2-3 days before grafting.
- Make sure that scions and rootstock have similar diameter of stem.
- Scion seedlings at a beginning of first true leaf stage. The true leaf size is 2-3 mm.
- Rootstock seedlings at a first true leaf stage. The long hypocotyls (7-9 cm) are desirable.
- The true leaf blade size is ~2 cm.
- Scalpel with handle works the best for this grafting method.
- Perforating tool. A plastic soldering tool works well. Alternatively you can create a tool by sharpening the edge of bamboo chopsticks (pencil sharpener works great).
- New trays filled with well moistened substrate.

Types of Grafting

Grafting methods involve such techniques as cleft grafting, tube grafting, whip grafting, tongue grafting, spliced grafting, flat grafting, saddle grafting, bud grafting, hole insertion grafting, and tongue approach grafting etc. These methods of grafting are briefly described as under:

Cleft grafting: For practicing this method of grafting, seeds of the rootstocks are sown 5-7 days earlier than those of the scion. The stem of the scion (at the four leaf stage) and the rootstock (at the 4-5 leaf stage) are cut at right angles, each with 2-3 leaves remaining on the stem. The stem of the scion is cut in a wedge and the tapered end fitted into a cleft cut in the end of the rootstock. The graft is held firm with a plastic clip.

Tube grafting: This method of grafting was developed by Lee *et al.* (1998). It makes possible to graft small plants grown in plug trays two or three times faster than the conventional method and is quite popular among Japanese seedling producers. Plants in small cells must be grafted at earlier growth stage and requires tubes with a smaller inside diameter. First the rootstock is cut at a slant. The scion is cut in the someway. Elastic tubes with side slit are placed onto the cut end of the rootstock. The cut ends of the scions are inserted into the tube, splicing the cut surfaces of the scions and root stocks together. While practicing the tube grafting in eggplant the seeds of *Solanum torvum* must be sown a few days earlier than those of the other rootstock species.

Method of preparing rootstock for grafting Method of preparing scion for grafting

Method of preparing grafts through cleft grafting

Tongue approach grafting: Melons and other cucurbitaceous plants are generally grafted by this method. It gives higher survival ratio because the root of the scion remains until the formation of the graft union. In this method, seeds of cu cumber are sown 10-13 days before grafting and pump kin seeds 7-10 days before grafting, to ensure uniformity in the diameter of the hypocotyl of the scion and rootstock. The shoot apex of the rootstock is removed so that the shoot cannot grow. The hypocotyl of the scion and rootstock are cut in such a way that they tongue into each other and the graft is secured with a plastic clip. The hypocotyl of the scion is left to heal for 34 days and then crushed between the fingers. The hypocotyl is cut off with a sharp razor blade three or four days after being crushed.

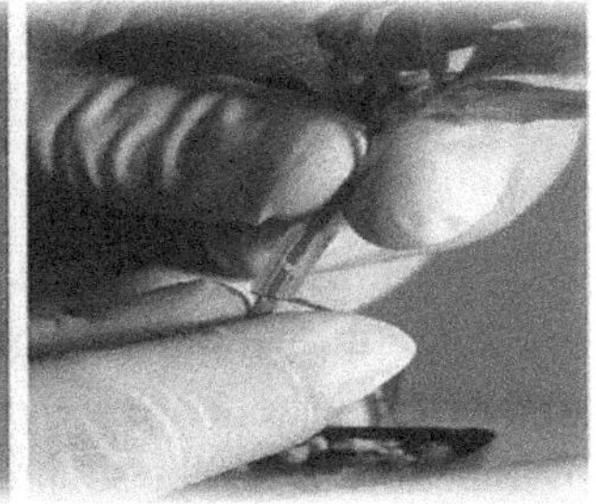

Method of preparing grafts through tube grafting

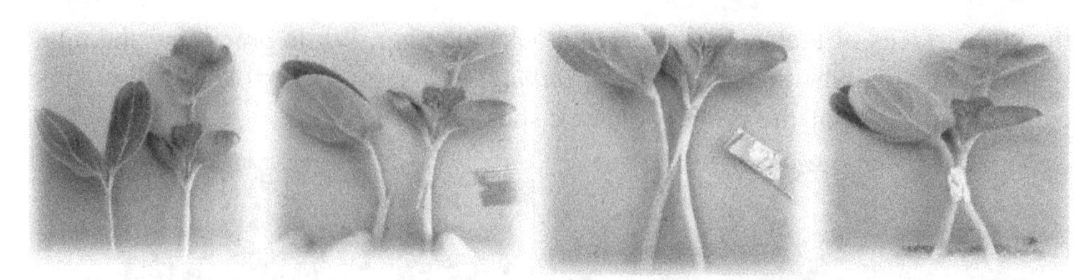

Method of preparing grafts through tongue approach grafting

Slant grafting: Recently this method of grafting has got popularity. It has been developed for robotic grafting. In this method, it is essential to remove the first leaf and lateral buds when a cotyledon of rootstock is cut on a slant.

Methodology of preparing grafts through Slant grafting

Hole-insertion grafting: This method is widely used for cucurbit crops. The rootstock leaf along with the growing point is removed and the scion is inserted into the stem of the rootstock. Following the five-step process, using a sharp utensil, such as a knife or razor blade, the scion is cut and then inserted into the upper portion of the rootstock. Rootstock seedlings should have one small true leaf and scion seedlings should have one or two true leaves. With a pointed probe, remove from the rootstock the true leaf along with the growing point. It is important to remove all of the growing point to prevent future shoot growth of the rootstock. This is one of the advantages of this type of graft. Use the probe to open a slit along one side on the upper portion of the rootstock's stem, where the stem connects to the cotyledons. Cut the scion and insert into the rootstock. Hold in place with a grafting clip. Place the grafted seedling in a chamber with high humidity at about 25°C and discard the unused parts.

Mechanized grafting: Grafting is arduous task and efforts are being made to reduce the labour required. Attempts have been made to mechanize grafting since 1987. There are several basic factors which govern the success of grafting by machine or robot such as seedling shape, location of cut, seedling gripping, cutting method, fixing materials and tools etc. Grafting robots for plug have been developed by combining the adhesive and grafting plants. This robot makes it possible for eight plugs of tomato, eggplant or pepper to be grafted simultaneously. Recently a fully automatic grafting system has been designed in which seedling quality estimation is clone by using fuzzy logic and neural network. Further, healing chamber with controlled atmospheric condition has also been designed to enhance the survival of grafts.

Method of preparing grafts through hole-insertion grafting

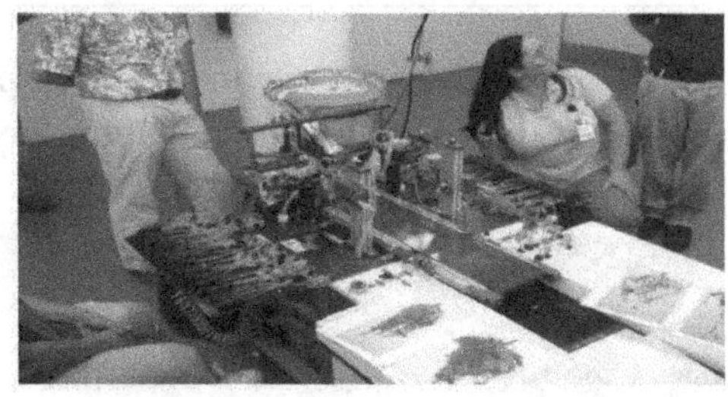

Use of grafting robot or machines for commercial production of grafts

Micro-grafting: Micro or In-vitro grafting is used to eliminate the viruses from infected plants using very small or micro-explants from meristematic tissues. But it is very expensive.

Important considerations for developing grafted vegetable plants

- Selection of root stock on the basis of purpose.
- Compatibility between rootstock and scions.
- Selection of grafting method

Problems Associated with Grafting

Various problems are commonly associated with grafting and cultivating seedlings. Major problems are the labour and techniques required for grafting operation and post graft handling of grafted seedlings for rapid healing.

Table 31.2 Problems associated with grafting and cultivating grafted vegetables seedling

Sr. No.	Factors	Category
1.	Labour	Grafting operation and post-graft care
2.	Techniques	Rootstocks
3.	Management	Fertilizer application
4.	Compatibility	Uneven senescence
5.	Growth	Excessive vegetative growth and physiological disorders
6.	Fruit quality	Size and shape, appearance, insipid taste, soluble solids, yellow band in flesh and internal decay
7.	Expense	Rootstock seeds
8.	Scion rooting	External rooting, internal or fused rooting

It is possible to develop wonder plants like 'Pomato' through grafting (A single plant will produce tomatoes as well as potatoes)

'Pomato'

Courtesy: CSK HPKV, Palampur

Potential Rootstocks

Rootstocks have a potential to provide tolerance against various abiotic and biotic stresses.

Table 31.3: Potential Rootstocks with special features of resistance against biotic and abiotic stresses

Crop	Species	Specific features
Tomato	*Solanum pennelli*	Tolerance to drought
	S. chessmanii	Resistant to salt
	S. galapagense	Tolerance to salt
	S. habrochaites	Resistance to cold as well as insects & diseases (TMV)
	S. chilense	Resistance to drought and diseases (CMV, TYLCV)
	S. neorickii	Resistant to bacterial diseases
	S. pimpinellifolium	Colour, quality, resistance to disease
	S. lycopersicum var. *cerasiforme*	Tolerance to humidity, resistance to fungi and root rot
	S. peruvianum	Resistance to tomato spotted wilt virus and RKN
Brinjal	*S. macrocarpon; S. gilo*	Tolerant to drought
	S. torvum	Resistance to *Verticillium* wilt, *Fusarium* wilt, RKN and tolerant to abiotic stresses.
	S. khasianum; S. viarum	Resistant to shoot and fruit borer (BSFB)
	S. xanthocarpum	Immune to phomopsis blight
	S. sisymbrifolium	Resistant to little leaf
	S. auriculatum	Immune to little leaf disease
	S. sisymbrifolium; S. indicum	Immune to RKN
Chilli	*C. chinensis; C. baccatum*	Anthracnose resistant species
	C. pubescens; C. microcarpum	Powdery mildew resistance species
Potato	*S. desmissum*	Resistant to late blight
Cucumber	*Cucumis hystrix*	Resistant to downy mildew, gummy stem blight, virus and nematode
Muskmelon	*Cucumis melo var. momordica*	Resistant to DM and PM
	Cucumis trigonus	Resistant to fruit fly
	C. anguria; C. ficifolia	Resistant to nematode
	C. metuliferus	
Pumpkin	*Cucurbita lundelliana*	Resistant to powdery mildew
Wax gourd	*Benincasa hispida*	Resistance to Fusarium wilt

32

MICROGREENS: A NEW CLASS OF EDIBLES FOR NUTRITION AND LIVELIHOOD

The spectrum of life in terms of income, life style and spending is changing rapidly with economic development. Diet related diseases such as obesity, diabetes, cardiovascular disease, hypertension, stroke and cancer are escalating both in developed and developing countries, in part due to imbalanced food consumption patterns. In developing countries like India, 13.5% people are chronically undernourished with Western-Asia and Sub-Saharan Africa, the most severely affected regions. Vegetables are oftenly referred to as Protective Food in view of nutritive and medicinal values and seve as one of the important components of Indian agriculture towards nutritional security of people. India is the 2nd largest producer across the globe occupying 10.32 million hectares area and producing 189.46 million tonne of vegetables but national food security is becoming a matter of increasing concern and poverty is reflected in the nutritional status of the people. The present per capita availability of vegetables in India is only 210 g against the requirement of 300g/ capita / day for normal health as per the Recommended Daily Allowance (RDA). Households in large cities in low-income countries like India spend 50-80 per cent of their incomes on food and nutritional deficits in macronutrients and essential micronutrients are common.

Now-a days, non-availability of fresh and pesticide residue free vegetables for consumption is increasingly becoming major concern for vegetarian population of our country. So, Microgreens: a new class of edible vegetables with lots of potential in term of nutritional ability to cure various deficiencies presents a homestead option towards nutritional security. Because, microgreens can easily be grown in urban or peri-urban areas either by specialized vegetable farmers or the consumers themselves, where land is often a limiting factor. Simultaneously, they also offer opportunities for rural population of our country to enhance dietary status of their food. Microgreens can easily be grown with and without soil organically in short period of span of 10-15 days around or inside residential areas. Moreover, microgreens are usually consumed raw, hence there is no loss or degradation of micronutrients through food processing. There are more than 25 microgreens commercially grown all over the world. Phytonutrient levels differ according to growth stages of the plant and often decrease from the seedling to the fully developed stage. Microgreens are 4-6 times more nutrient dense than their mature counterparts. So, microgreens can be termed as 'Functional Foods', which have health promoting or disease preventing properties. In recent years, consumption of microgreens has increased along with consumer awareness and appreciation for their tender texture, distinctive fresh flavours, vivid colours and concentrated bio-active compounds such as vitamins, minerals, antioxidants *etc.*

Historical Development

Historical importance of microgreens goes back to 1930s, when wheatgrass was grown dried and sold then sold as a medicine in most North American pharmacies. In the 1960s sunflower, buckwheat and radish were frequently grown as winter Greens, whereas during 1970s, healthy home grown "Grasses" were popularized for their health benefits. Chefs started growing "Cresses" and "Seedlings" for garnishing In the 1980s. The first documented use of the word "MICROGREENS" started in USA during 1998.Then in the 2000s, local producers throughout North America started distributing fresh "Microgreens" to their local retail outlets. Microgreens have started to appear at grocery stores so that food enthusiasts could enjoy them at home. In recent years, consumption of microgreens has increased along with consumer awareness and appreciation for their tender texture, distinctive fresh flavours and concentrated bio-active compounds such as vitamins, minerals, antioxidants as compared to mature leafy greens. Now, most popular chefs of India namely Sanjeev Kapoor, Vikas Khanna, Ranveer Brar, Kunal Kapoor are making blends of various microgreens in Indian food items and have now created interest in these items and put them in lime light amongst many people about their importance because of highly concentrated bio-active compounds.

Dozens of different types of vegetables can be planted to produce microgreens, with a range of tastes-mellow, spicy, tangy, earthy, nutty and crisp. Some of the most common varieties are things you may already grow to use the full plant: basil, parsley, cilantro, radishes, salad burnet, fennel, chervil, mustard, kale, collards, beets, pak choy and cabbage. Even carrots and beets can be sown and harvested as microgreens.

Microgreens: What are they?

Microgreens are a new class of edible vegetables, a very specific type which includes seedlings of edible vegetables, herbs or other plants, ranging in size from 5 to 10 cm. Microgreens contain three part central stem, cotyledon leaves and first pair of very young true leaves. Based on growth stages of plant, microgreens fall in the stage older than “Sprouts” and younger than “Babygreens”

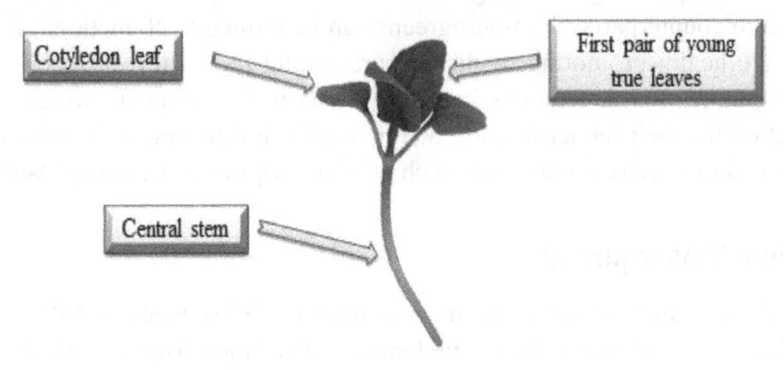

Diagrammatic representation of a microgreen

There are 70-80 types of microgreens grown worldwide but, ones listed in Table 32.1 have gained importance for homestead and commercial utility.

Table 32.1: List of important microgreens

Commercial name	Botanical Name	Family	Microgreen colour
Arugula	*Eruca sativa* Mill.	Brassicaceae	Green
Bull's blood beet	*Beta vulgaris* L.	Chenopodiaceae	Reddish green
Celery	*Apium graveolens* L.	Apiaceae	Green
Cilantro	*Coriandrum sativum* L.	Apiaceae	Green
Garnet amaranth	*Amaranthus hypochondriacus* L.	Amaranthaceae	Red
Golden pea tendrils	*Pisum sativum* L.	Fabaceae	Yellow
Green basil	*Ocimum basilicum* L.	Lamiaceae	Green
Green daikon radish	*Raphanus sativus* L.	Brassicaceae	Green
Magenta spinach	*Spinacia oleracea* L.	Chenopodiaceae	Red
Mizuna	*Brassica rapa* L.	Brassicaceae	Green
Opal basil	*Ocimum basilicum* L.	Lamiaceae	Greenish purple
Opal radish	*Raphanus sativus* L.	Brassicaceae	Greenish purple
Pea tendrils	*Pisum sativum* L.	Fabaceae	Green

[Table Contd.

Contd. Table]

Commercial name	Botanical Name	Family	Microgreen colour
Pepper cress	*Lepidium bonariense* L.	Brassicaceae	Green
Popcorn shoots	*Zea mays* L.	Poaceae	Yellow
Purple kohlrabi	*Brassica oleracea* L.	Brassicaceae	Purplish green
Purple mustard	*Brassica juncea* L.	Brassicaceae	Purplish green
Red beet	*Beta vulgaris* L.	Chenopodiaceae	Reddish green
Red cabbage	*Brassica oleracea* L.	Brassicaceae	Purplish green
Red mustard	*Brassica juncea* L.	Brassicaceae	Purplish green
Red orach	*Atriplex hortensis* L.	Chenopodiaceae	Red
Red sorrel	*Rumex acetosa* L.	Polygonaceae	Reddish green
Tartary buckwheat	*Fagopyrum tataricum* L.	Poaceae	Green

Source: Xiao *et al.*, (2012)

SELECTED RESEARCH WORK ON MICROGREENS

Nutritional Importance

Dagmar *et al.* (2010) carried out DPPH assay to estimate antioxidant activity as gallic acid equivalent in common and tartary buckwheat microgreens and observed higher antioxidant activity in tartary buckwheat than those of common buckwheat microgreens.Sudtirol-3was the only common buckwheat variety having higher antioxidant activity than cv. 01Z5100001 of tartary buckwheat.

Xiao *et al.* (2012) assessed commercially grown microgreens for different nutritional components and observed higher levels of TAA in cabbage, phylloquinone and violaxanthin in garnet amaranths, *á*-carotene and lutein in red sorrel, there by highlighting the relevance of individual microgreen for different nutritionally important compounds.

Ebert *et al.* (2014) evaluated four cultivars of amaranth separately as microgreen and mature greens for various nutritional parameters and found higher levels of neoxenthin, lutein, *á*-carotene and *â*-carotene in microgreens of cv. VI04470 compared to its fully grown stage. Likewise, cv. VI047164 emerged as single cultivar displaying higher level of chlorogenic acid compared to other microgreens as well as fully grown amaranths. The organoleptic study revealed high appreciation for appearance in cv. 047764, whereas'VI044470' and 'Hung-Shing-Tsai' received the highest rating for texture, taste and general acceptability.

There are some important vegetable crops which are gaining importance in some metro and big cities of India.

Red Amaranth	**Beetroot**	**Broccoli**	**Cress**
Sweet and tangy flavour, add vibrant dash colour to salad, used in garnishing.	Attractive, deep reddish metallic purple leaves, high antioxidant properties and rich in vitamins.	Rich in vitamins, minerals, enzymes, protein and chlorophyll, stimulate the immune system.	Peppery flavour, garnishing and addition to salads and sandwiches, source of vitamins A, C and S.
Dill	**Fenugreek**	**Kale**	**Linseed / Flaxseed**
Fine, feathery foliage and a great flavor, blends well with cucumbers, cheese, salmon and cabbage.	Rich in protein, vitamins A, D, E, B and minerals, stimulate the appetite and effective against anaemia and fatigue.	Cabbage like flavour, colourful leaves, add vibrancy to salads, antioxidants, prevent macula degeneration.	Spicy, tender, nutritious, rich in Omega-3 fatty acids, good source of vitamins, minerals, anti-oxidants and amino acids.
Radish	**Red Cabbage**	**Fennel**	**Mustard**
Rich in Ca, Fe, K, Zn, carotene, antioxidants, vitamins and protein, Stimulate immune system.	Rich in vitamins A, B, C, E, K, minerals and chlorophyll, stimulate immune system.	Higher in K, vitamins C, B and Phytonutrient, Decrease risk of heart disease.	High in antioxidants, protein, vitamins, and minerals, Stimulate blood circulation and effective against fever and cold.
Onion	**Pea**	**Red Veined Sorrel**	**Golden Corn**
Full of vitamins, minerals such as Ca, K, S, protein, enzymes and chlorophyll.	Nutritious and source of vitamins A, C, K and minerals Ca, Fe, Mg, P, K, amino acids and protein.	Boost eye sight, strengthen the immune system, build strong bones, prevent cancer, lower down blood pressure	Sweet flavour, used in garnishing, good source of vitamin B, antioxidant and carotenoids
Carrot			
Rich in β-carotene, phytonutrient like lutein and zeaxanthin, Good for beautiful skin, cancer prevention and anti-aging.			

Source: www.greenharvest.com.au

Pinto *et al.* (2015) carried out comparative analysis of mineral profiles in microgreens and mature greens of lettuce and observed significantly higher content of minerals like Ca, Mg, Fe, Mn, Zn, Cu, Mo and Se in microgreens compared to mature greens. The presence of lower level of nitrate content in microgreens than mature greens signifies their importance as safer food item in human diet particularly for children.

Xiao *et al.* (2016) analyzed mineral composition of popularly grown microgreens of Brassicaceae family and found variable potential of these microgreens with highest Ca, Mg in cauliflower, P in broccoli, K in arugula and Fe in kohlrabi purple. All the microgreens analyzed under study for the presence of heavy metal Cd were found free from this toxic element.

Germination

Lee *et al.* (2004) studied the effect of different seed treatments on seed germination of beet and chard and observed higher final germination percentage (FGT) of 99% and 91% in beet and chard, respectively through matric priming treatment. However, seed treatment with H_2O_2 registered earliest germination in both the types of *Beta vulgaris.* They further studied that horizontal vermiculite orientation resulted in a higher germination percentage along with longer radicles than vertical vermiculite orientation. Shoot fresh weight was little affected by vermiculite orientation in beet but was reduced slightly within the vertical orientation when seeds came from the bottom rather than top. Whereas, horizontal vermiculite orientation gave greater shoot fresh weight in chard than the vertical orientation.

Fertilizer Requirement

Kou *et al.* (2014) studied the effect of different sources of calcium on hypocotyl length of broccoli microgreens and observed significantly higher hypocotyl length with 10 mM $CaCl_2$ application. They further compared fresh weight, dry weight and calcium content in broccoli microgreens raised with water and best treatment (10 mM $CaCl_2$) and reported significantly higher value of these components in microgreens sprayed with $CaCl_2$.

Light Requirement

Brazaityte *et al.* (2015) carried out an experiment to determine the effect of UV-A on antioxidant content of beet microgreens under different lighting regimes with different degree of UV-A supplementation. The study revealed that supplemental UV-A resulted in increased DPPH radical scavenging activity. The lower UV-A irradiance level have no effect on phenol content but it increased significantly at +402 nm supplementation and anthocyanin content was also greatest under UV-A supplementation of +390 nm. Ascorbic acid significantly increased at +366 nm UV-A radiance which was further higher in +402 nm irradiance in EXP-2. At higher irradiance level, all supplemental UV-A had a positive effect on *á*-tocopherol content of microgreens with highest level in +366 nm. They also noticed greater effect of supplemental +366 nm irradiance on flavonol index. UV-A exposure increased nitrate content in microgreens, however maintained below the limit not affecting human health.

Viktorija and Akvile (2015) studied the effect of different light treatments on growth and nutritional parameters of mustard microgreens and observed no gain in hypocotyl length, plant height and leaf area in LED compared to control (HPS). However, artificial light treatment affected nutritional parameters significantly with highest impact of LED 250 treatment on ascorbic acid and total phenol content with at par response in LED 150. Whereas, total anthocyanin content was significantly higher in microgreens grown under LED 150 treatment. Artificial lighting also increased flavonol and $ABTS^+$ significantly over control irrespective of LED lighting.

Disease Management

Pill *et al.* (2011) studied the response of beet to damping off upon treatment with *Trichoderma harzianum* (Th) + *Trichoderma virens* (Tv) and observed lowest degree of damping off in seeds treated with *ThTv* at the rate of 1 mg per seed ball under 0.5 and 1.0 (*Pythium aphanidermatum*) levels of inoculum. However, seed treated with 0.25 mg recorded early germination.

Post Harvest

Chandra *et al.* (2012) studied the changes in off-odour of chinese cabbage treated with different sanitizers and packaging films during storage and observed minimum undesirable off-odour development in microgreens sprayed with citric acid followed by ethanol, when stored in polyethylene during storage time.

Kou *et al.* (2013) studied the effect of storage temperature on changes in aerobic mesophilic count and tissue electrolyte leakage of buckwheat microgreens and observed significantly higher bacterial count(1-2 log CFU/g) in microgreens stored at 15 and 20 °C than those stored at 10, 5, and 1 °C after 10 days of storage. Microgreens stored at 10°C showed minimum activity of bacteria even after 14 days of storage. Similarly, microgreens stored at lower temperatures revealed minimum electrolyte leakage and remained constant during all the period of storage.

The OTR level of 16.6 (m^2 s Pa) turned out to be best storage atmosphere for minimum electrolyte leakage, which was at par with the level of 29.5 (m^2 s Pa).

Looking to the importance of these edibles, Department of Vegetable Science, ASPEE College of Horticulture & Forestry, Navsari Agricultural University, Navsari, Gujarat, India also initiated experiments on effect of different light sources on growth and quality of different microgreens and concluded based on the performance of different microgreens under 3 different light sources namely incandescent (Tungsten light), fluorescent (tube light) and electroluminescent (LED light) that growth parameters like days to first harvest, leaf area (cm^2), fresh weight and quality parameters *viz*., ascorbic acid, β-carotene, N, P, K, Ca, total antioxidant activity and overall acceptability of all the microgreens under study (Fenugreek, Amaranthus, mustard, red cabbage, coriander, beet root, sweet corn) were evolved significantly under electroluminescent light is recommended for growing microgreens inside growing chamber/chamber/room. Fenugreek, beet root and red cabbage displayed significantly maximum ascorbic acid, N, Ca; *â*-carotene, K; antioxidant activity. Based on sensory evaluation, highest score for overall acceptability was obtained by Amaranth microgreens, which was followed by beet root and red cabbage microgreens.

Finally, electroluminescent light is recommended for growing microgreens inside growing chamber/chamber/room.

The Horticulture students of college are given skill development training on the component of microgreen cultivation under Student READY programme.

Microgreens: Growing Process

Now-a-days people are becoming aware about importance of microgreens. So, inhabitants in rural, urban area and peri-urban locations can utilize nutritional potential of microgreens at home and market. Although, growing process of microgreens is very easy but commercial basic requirements of microgreens it need to be taken into confederation for successful cultivation of microgreens. Some important requirements of some of the microgreens are given in Table 32.2.

Material and Media

Microgreens cultivation is not much costly because as they do not require much tool and material to grow. Selection of growing trays for the commercial cultivation should have good drainage capacity. Size of growing trays can appropriately be chosen depending upon availability of space and ease in handling and transportation of living microgreens. As far homestead cultivation is concerned, one can even use deposable trays for successful cultivation. Media should preferably be inert one like cocopeat, vermiculite and cocopeat alone or in combination of 3:1:1. As the concept of microgreens cultivation relates to provide pesticide free and nutritional rich food, so treatment with any chemical pesticides should be avoided. Moreover looking to vulnerability of microgreens particularly to damping-off disease, seeds can be treated with *Trichoderma harzianum and Trichoderma virens* alone or in combination (1 mg per seed ball). This practice is very important when trail is used as growing media (Pill *et al*., 2011).

Table 32.2: Growing preconditions for microgreens

Microgreen	Seed (g) / Tray (30 x 30 cm)	Soaking Time (h)	Depth of media Mix (cm)	Temp. (°C)	Maturity (Days)
Amaranth	2.5	NA	2	>22	16-25
Purple Basil	2.5	NA	1	>24	16-25
Beet root	12.5	24	2	16-25	16-25
Buckwheat	12.5	8-12	2	20-25	5-6
Cress	8	NA	1	16-25	5-14
Dill	5	NA	1	15-23	16-25
Kale	5	4-8	2	16-28	16-25
Linseed	36	NA	2	16-25	6-8
Mustard	2.5	8	2	16-25	15-20
Pea shoots	100-150	8-12	2	15-25	10-14
Radish	5	6-12	3	16-28	12
Cabbage	5	4-8	2	16-25	3-6
Arugula	3	NA	2	16-25	16-25
Sunflower	50	8-12	2	20-25	8-12

Source: www.greenharvest.com.au

Growing tray with small hole for better drainage

Tray filling by media

Desposable tray for homestead production

Sowing

One essential practice followed before the sowing or spreading of seeds in media is soaking. Seeds of spinach and fenugreek require soaking for getting good germination percentage (Table 2). Then seeds are sprinkled over the media with high density and covered with paper towel/ vermiculite/cocopeat. Generally, bigger sized of seeds are covered with vermiculite or cocopeat and smaller seeds with paper towel.

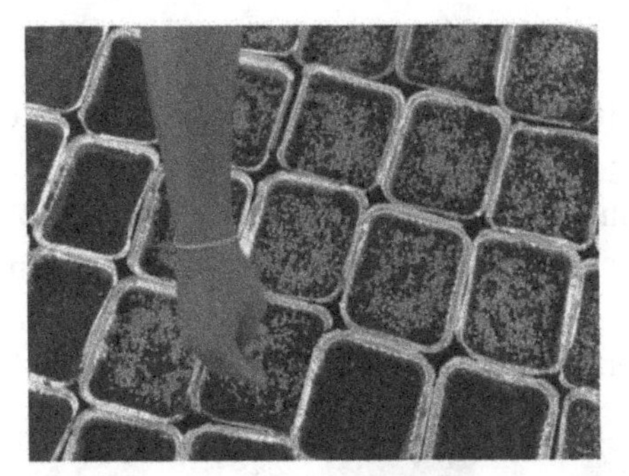

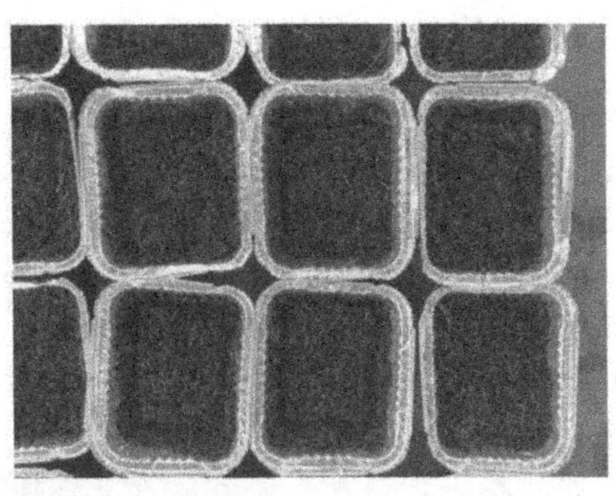

High density sowing of microgreens seeds

Aftercare

Though, microgreens do not require much care after the spreading of seeds, however sufficient moisture should be maintained through fine spray. High light requirement of 12-16 hours of period should preferably be maintained along with low humidity and good air circulation for better growth and development of microgreens.

Light watering with fine spray

Nutritional Requirement

As microgreens cultivation is aimed at to provide organic edible for better health of people. Fertilizer requirement for microgreens production is very minimum, which can easily be achieved through organic sources. Otherwise, one can also spray solution of 10 mM CaCl2 *i.e.* 1.10 g per 10 litres of water (Kou *et al.*, 2014).

Suplimental organic fertilizer spary by using fine spray

Harvesting

Microgreens are harvested at the appearance of 1st pair of true leaves. Most of the microgreens are ready for harvesting after the 10-15 days after sowing of seed. Microgreens cutting should be done above from media surface without roots. Some types of microgreens like coriander and fenugreek may regrow and can be cut several times. The media once used to growing the microgreens, can also be used successfully for another crop of microgreens.

Stage of Cutting and harvesting microgreens

Utility

In recent years, consumption of microgreens has increased along with consumer awareness and appreciation for their tender texture, distinctive fresh flavors and concentrated bioactive compounds such as vitamins, minerals, antioxidants as compared to mature leafy greens.

Homestead Utility

Microgreens are vivid in colour, so can be used in plate presentation and garnishing, which introduce hidden tangy flavours dishes. A tiny pile of microgreens can also be used to add flavours in salad. India represent vide variety of eatables in daily diet, nutritional enrichment of which can be done with microgreens.

Some of the homestead utilities of microgreens are presented below:

Blending of microgreens with different eatables

Commercial Utility

Microgreens are highly perishable in nature and it can't be stored for a long time in open as well as refrigerator conditions. To overcome this problem and to fulfill commercial utility of microgreens, they are sold as living microgreens. Living microgreens are the freshest and most nutritious greens and can be stored in the refrigerator for up to 14 days or at room temperature for 4-6 days with daily watering. They are sold in the market at rate of Rs. 200 per 100g. However, branding of microgreens is very important to make commercial utility successful so, proper packaging of living microgreens is essential to attract masses particularly new generation towards this new class of edible vegetable, as is being done in developed countries like USA.

Live microgreens being sold in USA

  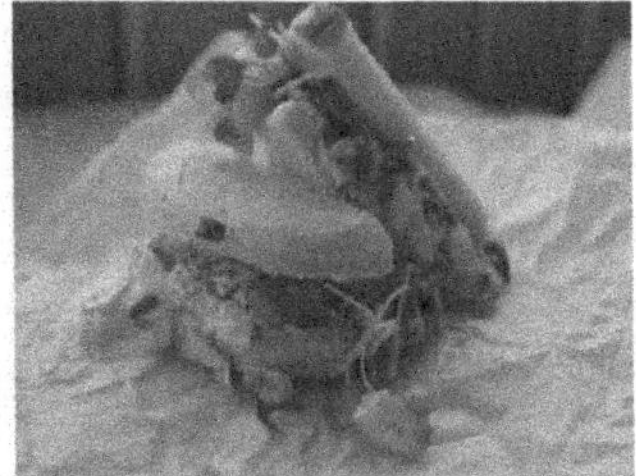

Commercial and homestead utility of microgreens

Microgreen Troubleshooting

Weak, skinny microgreens: The plants need more light compared their mature counterparts, otherwise microgreens may become weak and skinny.

Overcrowding: Excessive dense sowing may cause damping off, however it can be easily overcome by treating the media by Trichoderma.

Wrong sowing time: Some seeds may not germinate at very high or very low temperatures.

Over soaking: Over soaking of seeds may result in dead seeds.

REFERENCES

Abdelmageed, A.H.A., Gruda, N. and Geyer, B. 2004. Effects of temperature and grafting on the growth and development of tomato plants under controlled conditions. *In*: Deutsher Tropentag, Rural Poverty Reduction through Research for Development and Transformation; October 5-7. 2004, pp. 1-5.

Abdullahi, H.S., Mahieddine, F. And Sheriff, R.E. 2015. Technology impact on agricultural productivity: A review of precision agriculture using unmanned aerial vehicles. *In*: Wireless and Satellite Systems, 7th International Conference, WiSATS 2015. 6-7 July 2015, Bradford, UK. pp.388-400. Doi: 10.1007/978-3-319-25479-1_29.

Afroza, B., Wani, K.P., Khan, S.H., Jabeen, N., Hussain, K., Mufti, S. and Amit, A. 2010. Various technological interventions to meet vegetable production challenges in view of climate change. Asian Journal Horticulture. 5(2): 523-529.

Agele, S. and Cohen, S. 2009. Effect of genotype and graft type on hydraulic characteristics and water relations of grafted melon. Journal of Plant Interactions. 4(1): 59-66.

Agricultural and Processed Food Products Export Development Authority, New Delhi. https://apeda.gov.in.

Ahmad, P. and Prasad, M.N.V. 2012. Abiotic Stress Responses in Plants. Springer New York Dordrecht Heidelberg London. 465 p.

Albacete, A., Martinez-Andujar, C., Perez, A.M., Thompson, A.J., Dodd, I.C. and Alfocea, F.P. 2015. Unravelling rootstock x scion interactions to improve food security. Journal of Experimental Botany. 1-16.doi:10.1093/jxb/erv027.

Ali, M., Matsuzoe, N., Okubo, H. and Fujieda, K. 1992. Resistance of Non-tuberous Solanum to Root-knot Nematode. Journal of Japan Society of Horticultural Science. 60(4): 921-926.

Aloni, B., Cohen, R.,Karni, L.,Aktas, H. and Edelstein, M. 2010. Hormonal signaling in rootstock-scion interactions. Scientia Horticulturae. 127: 119-126.

Anil Kumar, Tiwari, G.N., Subodh Kumar and Mukesh Pandey. 2006. Role of greenhouse technology in agricultural engineering. International Journal of Agricultural Research, 1 (4): 364-372.

Anonymous 2018. Soilless culture for crop production, *In*: Hortimed. pp. 1- 48. http://www.just4growers.com/stream/growing-media/growing-mediahttps://vric.ucdavis.edu/pdf/hydroponics_soillessculture ofgreenhouse %20vegetables.pdf.

Anonymous 2020. Cost concepts - cost of cultivation and production. http:dacnet.nic.in/cost of cultivation.

Arupratan Singh. 2009. *Greenhouse Technology (The Future Concept of Horticulture)*. Kalyani Publishers. 320p.

Arvind Singh, Dhankhar, S.S. and Dashiya, K.K. 2015. Protected Cultivation of Horticultural Crops. pp. 1-8.10.13140/RG.2.218172.95364

Asada, K. 1999. The water-water cycle in chloroplasts: scavenging of active oxygens and dissipation of excess photons. Annual Review of Plant Biology. 50: 601-639.

Badgery-Parker, J. 1999. Light in the greenhouse. Agnote DPI/254 First edition, September 1999. https://www.dpi.nsw.gov.au/__data/assets/pdf_file/0007/119365/light-in-greenhouse.pdf.

Bagnaresi, P., Sala, T., Irdani, T., Scotto, C., Lamontarana, A., Beretta, M., Rotino, G.L., Sestili, L. and Sabatini, E. 2013. *Solanum torvum* responses to the root-knot nematode *Meloidogyne incognita*. BMC Genomics. 14: 540.

Bahadur, A., Rai, N., Kumar, R., Tiwari, S.K., Singh, A.K.,Rai, A.K., Singh, U., Patel, P.K., Tiwari, V. and Singh, B. 2015. Grafting tomato on eggplant as a potential tool to improve waterlogging tolerance in hybrid tomato. Vegetable Science. 42 (2): 82-87.

Baker, K.F., 1962. Principles of heat treatment of soil and planting material. Journal of the Australian Institute of Agricultural Science, 28: 118-126.

Bakker, N., Dubbeling, M., Gundel, S., Sabel-Koschella, U. and de Zeeuw, H. 2000. Growing cities, growing food. Urban agriculture on the policy agenda. Feldafing, Germany, Zentralstelle für Ernahrung und Landwirtschaft (ZEL), Food and Agriculture Development Centre.

Balass, M., Cohen, Y. and Bar-Joseph, M. 1992. Identification of a constitutive 45 kDa soluble protein associated with resistance to downy mildew in muskmelon (*Cucumis melo* L.) line PI 124111 F. Physiological and Molecular Plant Pathology. 41: 387-396.

Balint-Kurti, P.J., Dixon, M.S., Jones, D.A., Norcott, K.A. and Jones, J.D.G. 1994. RFLP linkage analysis of the *Cf-4* and *Cf-9* genes for resistance to *Cladosporium fulvum* in tomato. Theoretical and Applied Genetics. 88: 691-700.

Baudoin, W., Nono-Womdim, R., Lutaladio, N., Hodder, A., Castilla, N., Leonardi, C., De Pascale, S. and Qaryouti, M. 2013. *Good Agricultural Practices for Greenhouse Vegetable Crops: Principles for Mediterranean Climate Areas*. FAO Plant Production and Protection Paper 217. Food and Agriculture Organization of the United Nations, Rome, 2013. 603p. www.fao.org/publications.

Baudoin, W., Nono-Womdim, R., Lutaladio, N., Hodder, A., Castilla, N., Leonardi, C., De Pascale, S. and Qaryouti, M. 2013. Good Agricultural Practices for greenhouse vegetable crops: Principles for Mediterranean climate areas. *In*: FAO plant production and protection paper No. 217, Rome, 640p.

Beale, B. Vegetable Equipment and Irrigation Essentials. https://extension.umd.edu/sites/extension.umd.edu/files/_docs/programs/mdvegetables/Chap10-Veg-Equip-Irrigation-Essentials-web-version.pdf.

Benke, K. and Tomkins, B. 2017. Future food-production systems: vertical farming and controlled-environment agriculture, Sustainability: Science, Practice and Policy. 13 (1): 13-26. Doi: 10.1080/15487733.2017.1394054.

Bharathi, P.V.L. and Ravishankar, M. 2018.Technical partnership to support the green innovation centre for the agriculture and food sector (tomato value chain). *In*: Vegetable nursery and tomato seedling management guide for south and central India, World Vegetable Center, Taiwan. pp. 1- 40.

Bhatt, R.M., Laxman, R.H., Singh, T.H.. Divya, M.H., Srilakshmi and Nageswar Rao, A.D.D.V.S. 2014. Response of brinjal genotypes to drought and flooding stress. *Vegetable Science*, 41(2): 116-124.

B. Lathiya Jasmin, Sanjeev Kumar and Shivani Modi. 2018. Influence of Plant Growth Regulators on Growth and Yield of Greenhouse Tomato (*Solanum lycopersicum* L.). International Journal of Current Microbiology and Applied Sciences. 7 (8): 1603-1609.89.

Bletsos, F., Thanassoulopoulos, C. and Roupakias, D. 2003. Effect of grafting on growth, yield, and Verticillium wilt of eggplant. Horticulture Science. 38: 183-186.

Bletsos, F.A. 2006. Grafting and calcium cyanamide as alternatives to methyl bromide for greenhouse eggplant production. Scientia Horticulturae. 107: 325-331.

Boiteux L.S., Charchar J.M. 1996. Genetic resistance to rootknot nematode (*Meloidogyne javanica*) in eggplant (*Solanum melongena*). Plant Breeding. 115: 198-200.

Bolger, A., Scossa, F., Bolger, M.E., Lanz, C., Maumus, F., Tohge, T. Quesneville, H., Alseekh, S., Sorensen, I., Lichtenstein, G., Fich, E.A., Conte, M., Keller, H., Schneeberger, K., Schwacke, R., Ofner, I., Vrebalov, J., Xu, Y., Osorio, S., Aflitos, S.A., Schijlen, E., Jimenez-Gomez, J.M., Ryngajllo, M., Kimura, S., Kumar, R., Koenig, D., Headland, L.R., Maloof, J.N., Sinha, N., van Ham, R.C.H.J., Lankhorst, R.K., Mao, L., Vogel, A., Arsova, B., Panstruga, R., Fei, Z., Rose, J.K.C., Zamir, D., Carrari, F., Giovannoni, J.J., Weigel, D., Usadel, B. and Fernie, A.R. 2014. The genome of the stress-tolerant wild tomato species *Solanum pennellii*. Nature Genetics. 46(9): 1034-1039.

Bournival, B.L., Vallejos, C.E. and Scott, J.W. 1989. An isozyme marker for resistance to race 3 of *Fusarium oxysporum* f. sp. *lycopersici in tomato.* Theoretical and Applied Genetics. 78: 489-494.

Bradley, K. and Khosla, R. 2003. The Role of precision agriculture in cropping systems. Journal of Crop Production. 9 (1/2): 361-381

Brazaityte, A., Virsile, A., Jankaukiene, J., Sakalausiene, S., Samaulience, G., Sirtautas, R., Novickovas, A., Dabasinskas, L., Miliauskiene, J., Vastakuite, V., Bagdonaviciene, A. and Duckovskis, P. 2015. Intl. Agrophys. 29: 13-22.

Brown, JW. Light in the Greenhouse: How Much is Enough? http://www.cropking.com/articlelghe.

Cardina, J. and Doohan, D. J. Weed Biology and Precision Farming http://www.ipni.net/publication/ssmg.nsf/0/3954C7AF9D91293185257 9E500775FAC/$FILE/SSMG-25.pdf

Castilla, N. 2012. *Greenhouse Technology and Management*. 2nd Edition. CABI. 360p.

Chakrabarti, A.K. and Choudhury, B. 1975. Breeding brinjal resistant to little leaf disease. Proceedings of the National Academy of Sciences, India, Section B. 41: 379-385.

Chandra, D., Kim, J.G. and Kim, Y.P. 2012. Hort. Environ. Biotech. 53 (1): 32-40.

Chen, J.F., Moriarty, G. and Jahn, M. 2004. Some disease resistance tests in *Cucumis hystrix* and its progenies from inter-specific hybridization with cucumber. *In*: Lebeda, A., Paris, H.S. (Eds.). Progress in Cucurbit Genetics and Breeding Research. Olomouc: Palacky University, pp. 189-196.

Chia, R. Laser Land Levelling. The Global Magazine of Leica Geosystems. pp. 12-13. https://w3.leica-geosystems.com/media/new/product_solution/Leica_Geosystems_TruStory_Laser_Land_Levelling.pdf.

Chouraddi, M., Balikai, R.A., Mallapur, C.P. and Nayaka, P. 2011. Pest management strategies in precision farming. International Journal of Plant Protection. 4 (1): 227-230.

Christodoulakis, N.S., Lampri, P.N. and Fasseas, C. 2009. Structural and cytochemical investigaction of the silverleaf nightshade (*Solanum elaeagnifolium*), a drought-resistant alien weed of the Greek flora. Australian Journal of Botany. 57: 432-438.

CIAE Hand Held Vegetable Transplanter – Model – I (Single Row) http://www.ciae.nic.in

Colla, G., Fiorillo1, A., Cardarelli, M. and Rouphael, Y. 2014. Grafting to improve abiotic stress tolerance of fruit vegetables. Acta Horticulturae. 1041: 119-126.

Colla, G., Rouphael, Y., Rea, E. and Cardarelli, M. 2012. Grafting cucumber plants enhance tolerance to sodium chloride and sulphate salinization. Scientia Horticulturae. 135: 175-185.

D'Addabbo, T., Miccolis, V., Basile, M. and Candido, V. 2009. Chapter - 9: Soil Solarization and Sustainable Agriculture. pp. 218-274. (https://www.researchgate.net/publication/226312384).

Dagmar, J., Lenka, S. and Zdenek, S. 2010. Acta Agriculturae Slovenica. 95 (2): 157-162.

Danesh, D., Aarons, S., McGill, G.E. and Young, N.D. 1994. Genetic dissection of oligogenic resistance to bacterial wilt in tomato. Molecular Plant-Microbe Interactions. 7: 464-471.

Das, U., Pathak, P., Meena, M.K. and Mallikarjun, N. 2018. Precision farming a promising technology in horticulture: A Review. *International journal of pure and applied bioscience.* 6 (1): 1596-1606 (2018). doi: http://dx.doi.org/10.18782/2320-7051.3088.

Davis, A.R., Sakata, Y., Lopez-Galarza, S., Moroto, J.V. and Lee, S.G. 2008. Cucurbits grafting. Critical Reviews in Plant Sciences. 27 (1): 50-74.

de Souza, V.L. and Cafe-Filho, A.C. 2003. Resistance to *Leveillula taurica* in the genus Capsicum. Plant Pathology. 52: 613-619.

Direct Seeding Systems: Terms, Definitions and Explanations. http://www1.agric.gov.ab.ca/$department/deptdocs.nsf/all/agdex3483.

Dries, W. 2006. Design, Construction and Maintenance of Greenhouse Structures. *In*: Proc. IS on Greenhouses, Environmental Controls and In-house Mechanization for Crop Production in the Tropics and Sub-tropics, (Kamaruddin, R., Rukunuddin, I.H. and Hamid, N.R.A., Eds.). Acta Horticulturae, 710: 31-42.

Ebert, A.W., Wu, T.H. and Yang, R.Y. 2014. *In*: Sustaining Small Scale Vegetable Production and Marketing Systems for Food and Nutrition Security, 234-244 pp.

Ehret, D., Lau, A. Bittman, S. Lin, W. and Shelford, T. 2001. Automated monitoring of greenhouse crops. Agronomie, EDP Sciences. 21 (4): 403-414. https://hal.archives-ouvertes.fr/hal-00886117.

Evans, M.D., Dizdaroglu, M. and Cooke, M.S. 2004. Oxidative DNA damage and disease: induction, repair and significance. Mutation Research. 567(1): 1-61.

Farm Revolution -Sensors for Crop Pest Detection. http://blog.agrivi.com/post/farm-revolution-sensors-for-crop-pest-detection

Finkers, R., Bai, Y., van den Berg, P., van Berloo, R., Meijer-Dekens, F., ten Have, A., van Kan, J., Lindhout, P. and van Heusden, A.W. 2008. Quantitative resistance to *Botrytis cinerea* from *Solanum neorickii*. Euphytica 159: 83-92.

Food Safety and Good Practice Certification. http://www.fao.org/docrep/010/ag130e/AG130E12.htm

Forcella, F. Estimating the Timing of Weed Emergence. http://www.ipni.net/publication/ssmg.nsf/0/D26EC9A906F9B8C9852579E500773936/$FILE/SSMG-20.pdf.

Fry, W.E., Apple, A.E. and Bruhn, J.A. 1983. Evaluation of potato late blight forecasts modified to incorporate host-resistance and fungicide weathering. *Phytopathology*. 73(7):1054-1059. Doi: 10.1094/Phyto-73-1054

Galatti, F.S., Franco, A.J., Ito, L.A., Charlo, H.C.O., Gaion, L.A. and Braz, L.T. 2013. Rootstocks resistant to *Meloidogyne incognita* and compatibility of grafting in net melon. Ceres. 60: 432-436.

Gao, Q.H., Xu, K., Wang, X.F. and Wu, Y. 2008. Effect of grafting on cold tolerance in eggplant seedlings. Acta Horticulturae. 771: 167-174.

Garibaldi, A. and Gullino, M.L. 2010. Emerging soil-borne diseases of horticultural crops and new trends in their management. Acta Horticulturae. 883: 37-46.

Gass, I.G. 1960. The geology and mineral resources of the Dhali area. Memoir No. 4. Geological Survey Department, Nicosia, Cyprus.

Giacomelli, G.A., Sase, S., Cramer, R., Hoogeboom, J., MacKenzie, A., Parbst, K., Scarascia-Mugnozza, G., Selina, P., Sharp, D.A., Voogt, J.O., van Weel, P.A. and Mears, D. 2012. Greenhouse Production Systems for People. Proc. 28th IHC-IS on Greenhouse and Soilless Cultivation, Castilla, N., Ed.). Acta Horticulturae, 927: 23-38.

Giannakou, I.O. and Karpouzas, D.G. 2003. Evaluation of chemical and integrated strategies as alternatives to methyl bromide for the control of root-knot nematodes in Greece. Pest Management Science. 59: 883-892.

Gilardi, G., Gullino, M.L. and Garibaldi, A. 2011. Reaction of tomato rootstocks to selected soil-borne pathogens under artificial inoculation conditions. Acta Horticulturae. 914:345-348.

Gisbert, C., Prohens, J. and Nuez, F. 2011. Performance of eggplant grafted onto cultivated, wild, and hybrid materials of eggplant and tomato. International Journal of Plant Production. 5: 367-380.

Giuffrida, F., Cassaniti, C., Agnello, M. and Leonardi, C. 2015. Growth and ionic concentration of eggplant as influenced by rootstocks under saline conditions. Acta Horticulturae. 1086: 161-166.

Gousset, C., Collonnier, C., Mulya, K., Mariska, I., Rotino, G.L., Besse, P., Servaes, A. and Sihachakr, D. 2005. *Solanum torvum*, as a useful source of resistance against bacterial and fungal diseases for improvement of eggplant (*S. melongena* L.). Plant Science. 168(2): 319-327.

Greenhouse seeders. https://ag.umass.edu/greenhouse-floriculture/fact-sheets/greenhouse-s.

Greenhouse Technology. http://www.indiaeng.com/Kaveripakkam/04.

Guan, W. and Zhao, X. 2012. Defence mechanisms involved in disease resistance of grafted vegetables. Horticulture Science. 47(2): 164-170.

Hakkim, A.V.M, Joseph, A.E., Gokul, A.A.J. and Mufeedha K. 2016. Precision farming: The future of Indian agriculture. Journal of Applied *Biology* and *Biotechnology*. 4 (06): 068-072. Doi: 10.7324/JABB.2016.40609.

Hanson, P.M., Bernacchi, D., Green, S., Tanksley, S.D., Muniyappa, V., Padmaja, A.S., Chen, H., Kuo, G., Fang, D. and Chen, J. 2000. Mapping a wild tomato introgression associated with tomato yellow leaf curl virus resistance in a cultivated tomato line. Journal of American Society of Horticultural Sciences. 125: 15-20.

Harada, T. 2010. Grafting and RNA transport via phloem tissue in horticultural plants. Scientia Horticulturae. 125: 545-550.

Hartmann, H.T., Kester , D.E., Davies, F.T. and Geneve, R. 2010. Hartmann & Kester's Plant Propagation: Principles and Practices. Pearson Education, Inc., Publishing as Prentice Hall, One Lake Street, UpperSaddle River, New Jersey. pp. 415-464.

Hemming, S., van Henten, E., Van'r Ooster, B., Vanthoor, B. and Bakker, S. 2008. Systematic design of greenhouse crop production systems. Incose Insight. 4: 29-38. Doi: https://doi.org/10.1002/inst.200811429

Hemming, S., van Henten, E., van't Ooster, V., Vanthoot, B. and Bakker, S. Systematic design of greenhouse crop production systems. https://doi.org/10.1002/j.2334-5837.2008.tb00882.x.

http://aces.nmsu.edu/pubs/_circulars/CR556/.

http://agricoop.nic.in

http://atd.ztbl.com.pk/Art-LandLevling.aspx

http://dswcpunjab.gov.in/contents/data_folder/Laser_Level.htm

http://nhb.gov.in

http://nif.org.in/innovation/Onion_Transplanter/1.

http://software.ipni.net/article/SFT-7326

http://www.appropedia.org/Direct_seeding

http://www.cmeri.res.in/projects/pneumatic-precision-planter-vegetabl.

http://www.cropsreview.com/methods-of-planting.html

http://www.drishtiias.com/upsc-exam-gs-resources-Precision-farming1of914052018.

http://www.fao.org/hunger/en

http://www.fao.org/india/fao-in-india/india-at-a-glance/en/).

http://www.hydroponics.com.

http://www.indiaeng.com/Kaveripakkam/04

http://www.interurban.com/hydroponics

http://www.ncpahindia.com/application.php?appl_code=9.

http://www.ncpahindia.com/application.php?appl_code=9.

http://www.seeddynamics.com/seedtechnology/techniques/pelleting.

https://doi.org/10.1002/j.2334-5837.2008.tb00882.x

https://en.wikipedia.org/wiki/Soil_map.

https://shivkumardas.wordpress.com/agri-tech/laser-land-leveler/

https://www.dpi.nsw.gov.au/agriculture/horticulture/greenhouse/structu.

https://www.iec.cat/mapasols/Ang/Finalitat.asp?Grup=B&Opcio=7

https://www.irri.org

https://www.isro.gov.in

https://www.mapsofindia.com.

https://www.ncpahindia.com/precision-farming

https://www.ndsu.edu/pubweb/chiwonlee/plsc211/student papers/article.-3

https://www.prsindia.org/report-summaries/economic-survey-2018-19.

https://www.researchgate.net/publication/297543444

https://www.slideshare.net/mandalina/soilless-culture

https://www.incitecpivotfertilisers.com.au/~/media/Files/IPF/Documents/Agritopics/Gypsum.pdf

https://vric.ucdavis.edu/pdf/soil/ChangingpHinSoil.pdf

https://www.indiaagronet.com/indiaagronet/soil_management/CONT.

Humburg, D. Variable rate equipment- technology for weed control. http://www.ipni.net/publication/ssmg.nsf/0/F05D57E27B039458852579E5007671F1/$FILE/SSMG-07.pdf.

IIHR. 2015. Nematode management in protected cultivation. Technical Bulletin No. 48. (Rao, M. S., Umamaheswari, R., Chakravarthy, A. K., Manojkumar, R., Rajinikanth, R., Chaya, M. K., Priti K. and Narayanaswamy, B. Contributors). ICAR- Indian Institute of Horticultural Research, Hesaraghatta Lake Post, Bengaluru.

Imp Specialized vegetable production https://jhawkins54.typepad.com/files/equipment-for-specialized-vegetable-production-aly.pdf.

Ito, L.A., Charlo, H.C.O., Castoldi, R., Braz, L.T. and Camargo M. 2009. Rootstocks selection to gummy stem blight resistance and their effect on the yield of melon 'Bonus n° 2'. Revista Brasileira de Fruticultura. 31: 262-267.

Iyengar, K.S. Gahrotra, A., Mishra, A., Kaushal, K.K. and Dutt, M (Eds.). 2011. Greenhouse- A Reference Manual. National Comittee on Plasticulture Applications in Horticulture (NCPAH), NCPAH/TB/2010-11/13.

Johnson, S., Lnglis, D. and Miles, C. 2014. Grafting effects on eggplant growth, yield and verticillium wilt incidence. International Journal of Vegetable Science. 20 (1): 3-20.

Jurik, T.W., Weber, J.A. and Gates, D.M. 1984. Short-term effects of CO_2 on gas exchange of leaves of bigtooth aspen (*Populus grandidentata*) in the field. *Plant Physiology*. 75: 1022-1026.

Juvick, J.A., Bolkan, H. and Tanksley, S.D. 1991. The *Ve* gene for race 1 Verticillium resistance is on chromosome 7. Rep. Tomato Genet. Cooperative. 41: 23–24.

Kale, P.B., Mohod, U.V., Dod, V.N. and Thakare, H.S. 1986. Biochemical comparision in relation to resistance to shoot and fruit borer in brinjal. Vegetable Science. 13(2): 412-421.

Karlsson, M. 2016. Controlling the Greenhouse Environment. University of Alaska Fairbanks. 5-88/WV/11-16.

Katsoulas, N. and Kittas, C. 2008. Impact of greenhouse microclimate on plant growth and development with special reference to the Solanaceae. The European Journal of Plant Science and Biotechnology. 2 (Special Issue 1): 31-44.

Kienast-Brown, S., Libohova, Z. and Boettinger, J. Digital soil mapping. https://www.nrcs.usda.gov/wps/portal/nrcs/detail/soils/ref/?cid=nrcs14.

King, S.R., Davis, A.R., Liu, W. and Levi, A. 2008. Grafting for Disease Resistance. Horticulture Science. 43(6): 1673-1676.

King, S.R., Davis, A.R., Zhang, X. and Crosby, K. 2010. Genetics, breeding and selection of rootstocks for Solanaceae and Cucurbitaceae. Scientia Horticulturae. 127: 106-111.

Kou, L., Luo, Y., Yang, T., Xiao, Z., Turner, E.R., Lester, E.G. and Camp, M.J. 2013. Food Sci. Tech. 51: 71-78.

Kou, L., Yang, T., Liu, X., Haung, L. and Codling, E. 2014. Post Harvest Bio. Techno. 87: 70-78.

Krause, R.A., Massie, L.B. and Hyre, R.A. 1975. Blitecast: A computerized forecast of potato late blight. *Plant Disease Reporter*. 59(2): 95-98.

Kumar, P., Lucini, L., Rouphael, Y., Cardarelli, M., Kalunke, R.M. and Colla, G. 2015. Insight into the role of grafting and arbuscular Mycorrhiza on cadmium stress tolerance in tomato. Frontiers in Plant Science. 6: 477-496.

Kumar, S.; Patel, N. B. and Sravaiya, S. N. (2018). Influence of fertigation and training systems on yield and other horticultural traits in greenhouse cucumber. *Indian J. Hort.,* 75 (2): 252-258.

Kumaran, G.S. Innovations and interventions in horticultural mechanization. IIHR, Bengaluru. https://www.scribd.com/document/441355547/INNOVATIONS-OF HORTICULTURE-Dr-G-Senthil-IIHR-Bangalore-pdf.

Kusvuran, S., Kiran, S. and Ellialtioglu, S.S. 2016. Antioxidant enzyme activities and abiotic stress tolerance relationship in vegetable crops. Intech Journal. 481-506. doi.10.5772/62235.

Lee, J.M., Kubota, C., Tsao, S.J., Bie, Z., Echevarria, P.H., Morra, L. and Oda, M. 2010. Current status of vegetable grafting: diffusion, grafting techniques, automation. Scientia Horticulturae. 127: 93-105.

Lee, J.S., Pill, W.G., Cobb, B.B. and Olszewski, M. 2004. J. Hort. Sci. Biotech. 79 (4): 565-570.

Lee, S.H., Ahn, S.J., Im, Y.J., Cho, K., Chung, G.C., Cho, B.H. and Han, O. 2005. Differential impact of low temperature on fatty acid unsaturation and lipoxygenase activity in figleaf gourd and cucumber roots. Biochemical and Biophysical Research Communications. 330: 1194-1198.

Li, X., Eck, H.J.V., Voort, J.N.A.M.R.v.d., Huigen, D.J., Stam, P. and Jacobsen, E. 1998. Autotetraploids and genetic mapping using common AFLP markers: the *R2* allele conferring resistance to *Phytophthora infestans* mapped on potato chromosome 4. Theoretical and Applied Genetics. 96: 1121-1128.

Liaghat, S. and Balasundram, S.K. 2010. A Review: The role of remote sensing in precision agriculture. American Journal of Agricultural and Biological Sciences 5 (1): 50-55.

Liu, Y., Langemeier, M.R., Small, I.M., Joseph, L. and Fry, W.E. 2017. Risk management strategies using precision agriculture technology to manage potato late blight. Agronomy Journal. 109 (2): 562-575.

Lohan, S.K., Sidhu, H.S. and Singh, M. Laser Guided Land Leveling and Grading for Precision Farming. Available at: https://www.researchgate.net/publication/267641740. DOI: 10.13140/2.1.1103.9689.

Luck, J.D and Fulton, J.P. 2014. Precision Agriculture: Best management practices for collecting accurate yield data and avoiding errors during harvest. http://extensionpublications.unl.edu/assets/pdf/ec2004.pdf.

Luck, J.D. Yield Monitoring Systems: Understanding how we Estimate Yield. University of Nebraska. https://cropwatch.unl.edu/documents/Presentation_Yield%20Data%20Cleaning.pdf.

Luna, T. Wilkinson, K.M. and Dumroese, R.K. Seed Germination and Sowing Options. Tropical nursery manual. https://www.researchgate.net/publication/242270360.

Machineries for vegetable mechanization. http://ztmbpd.iari.res.in/technologies/equipments-implements/machiner.

Management of the Greenhouse Environment. https://www1.agric.gov.ab.ca/$department/deptdocs.nsf/all/opp2902.

Mandal, D. and Ghosh, S.K. 2000. Precision farming: The emerging concept of agriculture for today and tomorrow. Current Science. 79 (12): 1644-1647.

Mandal, S.K. and Maity, A. 2013. Precision farming for small agricultural farm: Indian scenario. American Journal of Experimental Agriculture. 3 (1): 200-217.

Mishra, A. Sundaramoorthi, K., Chidambara, R. P. and Balaji, D. 2003. Operationalization of Precision Farming in India. *In*: Map India Conference 2003. © GISdevelopment.net.

Model Bankable Project on Protected Cultivation in Haryana (www.nhm.nic.in).

Mondal, P. and Basu, P. 2009. Adoption of precision agriculture technologies in India and in some developing countries: Scope, present status and strategies. Progress in Natural Science. 19: 659-666. Doi:10.1016/j.pnsc.2008.07.020.

Nandede, B.M., Carpenter, G., Byale, N.A., Rudragouda, C., Jadhav, M.L. and Pagare, V. 2017. Manually operated single row vegetable transplanter for vegetable seedlings. International Journal of Agriculture Sciences. 9 (53): 4911-4914.

Natikar, P.K., Balikai, R.A. and Anusha, C. 2016. Pest management strategies in precision farming. Journal of Experimental Zoology. 19 (1): 1-8.

Naved Sabir and Balraj Singh. 2013. Protected cultivation of vegetables in global arena: A review. Indian Journal of Agricultural Sciences, 83 (2): 123–35.

Nishtha Patel, Sanjeev Kumar, R.V. Tank, Priyanka Patel and J.N. Tiwari. 2016. Response of tomato (*Solanum lycopersicum* L.) to varying levels of spacing and training under protected conditions. Green Farming. 7 (2): 438-441.

P.M. Chauhan, P.M. 2018. Agrovoltaic: A novel technology for doubling the income of farmers. *In*: Technologies and Sustainability of Protected Cultivation for

Hi-Valued Vegetable Crops (Sanjeev Kumar, Patel, B.N., Saravaiya, S.N. and Patel, N.B., Eds.). Navsari Agricultural University, Navsari. pp. 215-230.

Pardossi, A., Carmassi, G., Diara, C., Incrocci, L., Maggini, R. and Massa, D. 2011. *In*: Fertigation and substrate management in closed soilless culture, Pisa, Italy, pp. 1-64.

Parish, R.L. 2005. Current developments in seeders and transplanters for vegetable crops. HortTechnology. 15 (2): 346-351.

Patil, S.S. and Bhalerao, S.A. 2013. Precision farming: The most scientific and modern approach to sustainable agriculture. *International Research Journal of Science and Engineering.* 1 (2): 21-30.

Pill, W.G., Colling, C.M., Gregory, N. and Evans, T.A. 2011. Scientia Horticulturae. 129: 914-918.

Pinto, E., Almeida, A.A., Aguiar, A.A. and Ferreira, I.M.P.L.V.O. 2015. J. Food Comp. Anal. 37: 38-43.

Plant Response to Carbon Dioxide (CO_2) Enrichment. http://www.gardenandgreenhouse.net/articles/october-2013/plant-resp

Poudel, M. and Dunn, B. 2017. Greenhouse carbon dioxide supplementation. Oklahoma Cooperative Extension Fact Sheets. http://osufacts.okstate.edu.

Radha Manohar, K. and Igathinathane, C. 2019. *Greenhouse Technology and Management.* B.S. Publications, Hyderabad. 234p.

Rains, G.C. and Thomas, D.L. Precision Farming- An Introduction. University of Georgia Research And Extension Personnel. https://www.researchgate.net/publication/277791290

Reddy P.P.. 2003. *Protected Cultivation.* Springer Publications. USA.

Reddy, P.P. 2011. *Sustainable Crop Protection under Protected Cultivation.* Springer, Publications. USA.

Responses to Temperature & Light - Greenhouse Grower. http://www.greenhousegrower.com/production/plant-culture/responses-...1.

Richards, M. B., Bahl, K. B., Jat, M. I., Lipinski, B., Monasterio, I. O. and Sapkota, T. 2016. Site-Speciûc Nutrient Management: Implementation guidance for policymakers and investors. Practice brief on CSA. Doi: 10.13140/RG.2.1.1573.5448.

Robert, M.W. 1992. *An Introduction to Greenhouse Production.* Ohio Agricultural Education Curriculum Materials Service. The Ohio State University Ohio.

Rocha, J.V.. Precision Farming and Geographic Systems. https://www.researchgate.net/publication/37680195_GIS_and_remote_sensing_application_on_precision_agriculture?

Roger, H. 2016. Carbon Dioxide Enrichment Methods. https://www.hydrofarm.com/resources/articles/co2_enrichment.php.

Rojas, R.V. 2012. Digital soil mapping. http://www.fao.org/fileadmin/user_upload/GSP/docs/Presentation_NEMA_Inception/GSP_03_April_DSM.pdf.

Sabir, N., Singh, B., Hasan, M., Sumitha, R., Sikha Deka, Tanwar, R.K., Ahuja, D.B., Tomar, B.S., Bambawale, O.M. and Khan, E.M. 2010. Good Agricultural Practices (GAP) for IPM in Protected Cultivation, Technical Bulletin No. 23. National Centre for Integrated Pest Management, New Delhi, India, p. 16.

Sahoo, R.N. 2003. Precision Farming : Concept and Approaches. In: Proceeding of DOS funded 12th Winter School on Remote Sensing Application in Agriculture with special emphasis on Precision Farming. pp. 1-10.

Saravaiya, S.N. Patel, N.B. and Sanjeev Kumar. 2014. Protected Cultivation: Future Technology for Vegetable Crops. *In*: Global Conference on Climate Smart Horticulture, May 28-31, 2014. 260-277pp.

Sanjeev Kumar, Patel, N.B. Saravaiya, S.N. and Desai, K.D. 2015. Economic viability of cucumber cultivation under NVPH. African Journal of Agricultural Research. 10 (8): 742-747.

Sanjeev Kumar and Nishtha Patel. 2016. Effect of spacing and supportive training on quality parameters in tomato under protected culture. Bioinfolet. 13 (2 B): 420 – 424.

Sanjeev Kumar, Chaudhari Varsha I., Saravaiya, S.N. and Dev Raj. 2017. Potentiality of greenhouse cucumber cultivars for economic and nutritional realization. International Journal of Farm Sciences. 7 (1): 1-7.

Sanjeev Kumar, Patel, B.N., Saravaiya, S.N. and Patel, N.B. (Eds). 2018. Technologies and Sustainability of Protected Cultivation for Hi-Valued Vegetable Crops. Navsari Agricultural University, Navsari, Gujarat, India. February 01-03, 2018. 494p.

Sanjeev Kumar, Patel, N.B. and Saravaiya, S.N. 2018. Influence of fertigation and training systems on yield and other horticultural traits in greenhouse cucumber. Indian Journal of Horticulture, 75(2): 252-258.

Sanjeev Kumar, Patel, N.B. and Saravaiya, S.N. 2019. Studies on *Solanum torvum* Swartz rootstock on cultivated eggplant under excess moisture stress. Bangladesh Journal of Botany. 48 (2): 297-306. Doi: 10.3329/bjb.v48i2.47671.

Sanjeev Kumar, N.B. Patel and S.N. Saravaiya. 2020. Pruned side shoots as planting material: Opening new dimensions for sustainable greenhouse cucumber production system. Indian Journal of Horticulture. 77 (2): 307-314.

Sanjeev Kumar, Chaudhari Varsha I., Saravaiya, S.N. and Dev Raj. 2017. Potentiality of greenhouse cucumber cultivars for economic and nutritional realization. International Journal of Farm Sciences. 7 (1): 1-7.

Sanjeev Kumar, Lathiya Jasmin B. and S.N. Saravaiya, S.N. 2018. Microgreens: A new beginning towards nutrition and livelihood in urban-peri-urban and rural continuum. *In*: Technologies and Sustainability of Protected Cultivation for Hi-Valued Vegetable Crops (Sanjeev Kumar, Patel, B.N., Saravaiya, S.N. and Patel, N.B., Eds.). Navsari Agricultural University, Navsari. pp. 246-261.

Sanjeev Kumar, Narendra Singh and D.J. Chaudhari. 2018. Profitability of Capsicum Cultivation under Protected Condition. Chemical Science Review and Letters. 7 (28): 900-904.

Sanjeev Kumar, Nikki Bharti and Saravaiya, S.N. 2018. Vegetable Grafting: A Surgical Approach to combat biotic and abiotic stresses- A review. Agricultural Reviews. 39(1): 1-11.

Sanjeev Kumar, Patel, N.B. and Saravaiya, S.N. 2018. Analysis of bell pepper (*Capsicum annuum*) cultivation in response to fertigation and training systems under protected environment. Indian Journal of Agricultural Sciences. 88 (7): 1077-82.

Saravaiya, S.N.; Sanjeev Kumar, Vadodaria; J.R., Verma, P. and Varma, L. R. 2016. Vertical Farming: An Alternative Farming System for Future. *In*: Workshop Compendium on Forward Thinking for Agricultural Development in Western India, February 8-10, 2016, SDAU, Sardarkrushinagar. pp. 57-61.

Scheffe, K. and McVey, S. Soil mapping concepts. https://www.nrcs.usda.gov/wps/portal/nrcs/detail/soils/ref/?cid=nrcs14.

Sengupta, A. and Banerjee, H. 2012. Soil-less culture in modern agriculture. World Journal of Science and Technology. 2 (7): 103-108.

Shamshiri, R.R., Kalantari, F., Ting, K. C., Thorp, K.R., Hameed, I.A., Weltzien, C., Ahmad, D. and Shad, Z.M. 2018. Advances in greenhouse automation and controlled environment agriculture: A transition to plant factories and urban agriculture. International Journal of Agricultural and Biological Engineering. 11 (1): 1-22. Doi: 10.25165/j.ijabe.20181101.3210.

Si, Y., Liu, G., Lin J., Lv, Q. and Juan, F. 2008. Design of control system of laser leveling machine based on fussy control theory. *In*: Computer And Computing

Technologies in Agriculture, Volume II (Li, D., Ed.). CCTA 2007. The International Federation for Information Processing, Springer, Boston, MA. pp. 1121-1127

Singh, A. Dhankhar, S. S. and Dahiya, K. K. (2015). In: Training manual: Protected Cultivation of Horticultural Crops, CCS Haryana Agricultural University, Hissar, pp.: 1-88. (http://hau.ernet.in/).

Singh, A. Dhankhar, S.S. and Dahiya, K.K. 2015. *In*: Training manual: Protected Cultivation of Horticultural Crops, CCS Haryana Agricultural University, Hissar. pp. 1-88. http://hau.ernet.in/.

Singh, A.K. Precision Farming. http://apps.iasri.res.in/ebook/EBADAT/6-Other%20Useful%20Techniques/14-Precision%20Farming%20Lecture.pdf

Srinivasan, A. Relevance of Precision Farming Technologies to Sustainable Agriculture in Asia and the Pacific. http://www.sristi.org/mtsa/s_7_4.htm

Stafford, J.V. 2000. Implementing Precision Agriculture in the 21st Century. Journal of *Agricultural Engineering Research.* 76: 267-275. Doi: 10.1006/jaer.2000.0577.

Sudduth, K.A. Engineering Technologies for Precision Farming. https://www.semanticscholar.org/paper/ENGINEERING-TECHNOLOGIES-FOR-PRECISION-FARMING-Sudduth/c81f739bf0208f20a85bc58446c811931 1b7d8cb#citing-papers.

Techniques of Propagation by Seed. https://aggie-horticulture.tamu.edu/faculty/davies/pdf%20stuff/ph%20final%20galley/Chap%208%20-%20M08_DAVI4493_08_SE_C08.pdf.

TNAU Agritech Portal: Good Agricultural Practices (GAP). http://agritech.tnau.ac.in/gap_gmp_glp/gap_about.html

TNAU Tractor mounted three row plug type vegetable transplanter. Extension Bulletin No.CIAE/FIM/2006/64. https://aicrp.icar.gov.in/fim/wp-content/uploads/2017/01/30.-TNAU-Tractor-mounted-plug-type-vegetable-transplanter.pdf.

Tran, D.V. and Nguyen, N.V. The concept and implementation of precision farming and rice integrated crop management systems for sustainable production in the twenty-first century.https://pdfs.semanticscholar.org/f67d/28db59c04514336f8b16860c7e25dd4e5e5f.pdf?_ga=2.228042840.454061046.1590062449-444517366.1512281757. http://agropedia.iitk.ac.in/content/precision-agriculture-glance.

Tsuga, K. 2000. Development of Fully Automatic Vegetable Transplanter. Japan Agricultural Research Quarterly. 34: 21- 28.

van Henten, E.J. 2006. Greenhouse mechanization: State of the art and future perspective. Acta Horticulturae. 710: 55-70.

Van Os, E., Gieling, T.H., and Lieth, H.H. (2008). Technical Equipment in Soilless Production Systems. *In*: Soilless Culture Theory and Practice (Raviv, M. and Lieth, J.H., Eds.). Elsevier Publications. pp. 147-207. Doi: https://doi.org/10.1016/B978-044452975-6.50007-1.

Viktorija, V. and Akvile, V. 2015. Food Sci., 1: 111-117.

Wagh, R.M. 2017. Precision Farming, Remote sensing, geographical information system: a new paradigm for agricultural production in India. International Archive of Applied Sciences and Technology. 8 (4): 04-09. Doi: .10.15515/iaast.0976-4828.8.4.49.

Why direct seed? http://www.directseed.org/

Wittwer, S.H. and Castilla, N. 1995. Protected cultivation of Horticultural crops worldwide. HortTech. 5 (1): 6-23.

www.globalgap.org

www.greenharvest.com.au

www.ipni.net/publication/ssmg.nsf

Xiao, Z., Codling, E.E., Luo, Y., Nou, X. and Lester, E.G. 2016. J. Food Comp. Anal. 49: 87-93.

Xiao, Z., Lester, E.G., Luo, Y. and Wang, Q. 2012. J. Agric. Food Chem. 60: 7644-7651.

Yadav, H.K., Vijay Kumar and Yadav, V.K. 2015. Potential of solar energy in India: A Review. International Advanced Research Journal in Science, Engineering and Technology. 2 (1): 63-66. Doi: 10.17148/IARJSET.

Yield Monitoring and Mapping. Crop Watch. https://cropwatch.unl.edu/ssm/mapping.

Zhang, C. and Kovacs, J.M. 2012. The application of small unmanned aerial systems for precision agriculture: a Review. Precision Agriculture: An International Journal on Advances in Precision Agriculture. 13: 693-712. Doi: 10.1007/s11119-012-9274-5.

Zhou, S. and Zong, L. 2016. Research of wireless sensor monitoring network of melon fly under different temperatures and other environmental conditions. *International Journal Bioautomation*. 20(1): 125-134.

Zude-Sasse, M., Fountas, S., Gemtos, T.A. and Abu-Khalaf, N. 2016. Applications of precision agriculture in horticultural crops. *European Journal of Horticultural Science* 81(2): 78-90. Doi: http://dx.doi.org/10.17660/eJHS.2016/81.2.2.